W9-BAG-508

3 1163 00096 1424

9/98

DATE DUE

RIVERDALE

DATE DUE
10 17 01

DEMCO 128-8155

RIVERDALE PUBLIC LIBRARY DISTRICT
208 West 144th Street
Riverdale, IL 60827-2788
Phone: (708) 841-3311

SUPERCROSS.

Joe Bonnello

Motorbooks International
Publishers & Wholesalers ®

RIVERDALE PUBLIC LIBRARY DISTRICT

J 796.7
BON

Dedication

In memory of Donny Schmit
1990 125cc World Motocross Champion
1992 250cc World Motocross Champion
1968–96

First published in 1997 by Motorbooks International Publishers & Wholesalers, 729 Prospect Avenue, PO Box 1, Osceola, WI 54020-0001 USA

© Joe Bonnello, 1997

All rights reserved. With the exception of quoting brief passages for the purpose of review no part of this publication may be reproduced without prior written permission from the Publisher

Motorbooks International is a certified trademark, registered with the United States Patent Office

The information in this book is true and complete to the best of our knowledge. All recommendations are made without any guarantee on the part of the author or Publisher, who also disclaim any liability incurred in connection with the use of this data or specific details

We recognize that some words, model names and designations, for example, mentioned herein are the property of the trademark holder. We use them for identification purposes only. This is not an official publication

Motorbooks International books are also available at discounts in bulk quantity for industrial or sales-promotional use. For details write to Special Sales Manager at the Publisher's address

Library of Congress Cataloging-in-Publication Data Available

On the front cover: Jeremy McGrath styles at the 1996 Paris Supercross. Americans who run the European supercross races held in the off-season usually don't have the luxury of racing their own works bikes. Most use production bikes with works pipes, silencers, suspension components, and team-issue graphics and plastic. In 1996, the factory Honda riders were on Honda of Troy-badged machines.

On the frontispiece: From these carefully drawn plans, a crew of heavy equipment operators will create a supercross track that will live for only a few short days.

On the title page: Donny Schmit, celebrating one of his two World Championships. After winning those European titles in 1990 and 1992, Schmit returned to the United States to go into semi-retirement, racing four-strokes and in occasional Outdoor Nationals. Born in 1968 in Bloomington, Minnesota, Schmit died suddenly from a rare disease in 1996.

On the back cover: Casey Johnson on the SplitFire Kawasaki KX125. The chromed plastic was flashy but fragile. The coating chipped and cracked easily, and had to replaced after only a few races.

ISBN-0-7603-0320-7

Printed in Hong Kong

I. title
X. motorcross

CONTENTS

THE HISTORY OF SUPERCROSS
By Eric Johnson

Supercross is the prodigal son of motocross racing, an American phenomena born in 1972 that has grown to become the world's most popular form of off-road motorcycle racing. At that time, motocross racing was dominated by European riders, and the sport was just beginning to take off in the United States. Motocross would become wildly popular in the States, however, and American riders would become the best in the world by the mid-1980s. But in the early 1970s, the concept of building a small motocross track in the center of a giant stadium was a bold leap.

The first design for this spin-off sport was a rugged outdoor race track assembled on the infield of the Daytona International Raceway, initiated by Gary Bailey, but the conversion of ballparks to motocross tracks was conceived by Mike Goodwin.

Current American Motorcyclist Association (AMA) motocross and supercross referee Ron Crandall has been involved with supercross since the sport's conception in 1972. For that first race held in the Los Angeles Coliseum, Crandall was working as Goodwin's right-hand man in creating and directing both track construction and event logistics. He saw the entire process come together first hand.

"Mike Goodwin drew a sketch of the Coliseum and took it to two good bulldozer operators and asked them point blank, 'Can you guys do this kind of thing?' They said, 'Sure!' Then he went over to Vic Wilson at Saddleback Park to get a read from another voice of reason," Crandall said.

Bob Hannah burst onto the scene in 1977, a time when supercross was making the transition to the big time. A long-time Yamaha rider, the "Hurricane" switched to Honda later in his career. *Paul Buckley*

Rick Johnson in action on a CR500 at the USGP in 1990.

Legend has it that the original sketch of the track design for that early track was done on a cocktail napkin. According to Crandall, the legend is true. "The management of the Los Angeles Coliseum at that particular time—I know one of them pretty well—told me that he was having dinner and 'Mike had one too many cocktails and by golly, this is what he drew,'" Crandall recalls.

Goodwin's idea may have been gleaned from a race held in early March of 1972, within the infield of the Daytona International Speedway. Daytona, the crown jewel of Bill France's NASCAR empire, was looking for a support race to its highly successful Daytona 200 roadrace. East Coast motocross veteran Gary Bailey was commissioned to build a long, rough, bike-breaking circuit that was etched out of the grassy section of the Daytona's tri-oval infield. By all accounts, the Daytona track was similar to an outdoor national circuit, and it remains that way today.

Daytona may have been the catalyst, but it was Goodwin's vision to make the event a show. "As far as history goes, you have to give some credit to Bailey for doing it in Daytona, putting it in there first," claims Crandall, "but Mike was the first one to step out of the line and lay his money down and put it in a true stadium. He had some great backing from two rock promoters who were in with him."

"Goodwin's ideas were pure and simple." Crandall goes on to say, "He wanted to get peo-

ple in, let them watch something really exciting, and not get them dirty and dusty."

The initial Coliseum stadium track was like anything the first time out, a total trial-and-error affair. "The AMA (American Motorcyclist Association) flew out a guy named John 'Lightbrown' Lincoln to do the first one because no one on the West Coast would even touch it because they just didn't understand the concept," said Crandall. "Everybody was pretty skeptical at first. In fact, I remember at one point in time, Goodwin came running down the stadium steps looking like all the wild man and yelled 'What are you guys building, a golf course? We are not going to putt out here, we're going to race motorcycles, and I want the damn thing rough." The dozers were sent back out, and the track got rougher.

Soon thereafter, the riders took to the funky new circuit in an effort to get a feel of things. Despite some initial negativity, everybody seemed amiable to giving the new circuit a chance. Said Crandall, "Complaints were minimal. It was the very beginning, so everybody was excited about trying to do something about furthering the sport."

Later that historic Saturday evening, 23,000 spectators came out to watch a 16-year-old San Diego-based racer named Marty Tripes ride away with the victory. America had just witnessed its very first supercross event.

When the smoke, screaming, and dust settled, a number of people had big smiles on their faces.

"Everybody was talking about how awesome it was," said Crandall. "They were saying, 'What a concept. It's about time someone made it comfortable to watch a good motorcycle race by putting it into an entertainment establishment.'" The experiment worked, and a new sport was born.

Other promoters soon became involved; the motorcycle manufacturers jumped on board and sponsors such as Coca-Cola and the big beer companies started to sniff around, and supercross began to evolve rather quickly. In 1974, events were added in Houston, Texas, and an official AMA-sanctioned supercross series was formed.

After running both a 250cc and 500cc class in 1974 and 1975, the AMA and the

Ron Lechien was one of the most naturally talented riders of his time. He won his first supercross at age 16. *Paul Buckley*

manufacturers decided to combine the two divisions into a single, 250cc-only championship. In addition, two new venues would be added to the schedule, one in Pontiac, Michigan, and the other in Anaheim, California.

During the next few years, supercross transformed into a serious, professional sport. Said Crandall of the era, "Things were very crude at first, then they really started to develop around 1977. . . The tracks were also improving. During this time, and the years to follow, Semics, Kars-

makers, Gaylon Mosier, and Broc Glover would spend an extra hour and a half, or two hours with me walking around the track, talking about things, and taking notes. Between all of us, we put together a diagram that would produce a good supercross track."

The emergence of supercross as a big-time sport also saw a new legend appear. Bob Hannah took the supercross circuit by storm, reeling off six wins during the 1977 season's 10-race series. "Hurricane" Hannah would go on to

The Los Angeles Coliseum parastyle jump is one of the most famous obstacles in supercross history. The giant big-screen television in the background showed live coverage as the riders briefly left the stadium, as well as instant replays of important happenings.

completely dominate the stadiums, marching on to claim the series championship in both 1978 and 1979. Possessing an almost demonic-like work ethic, Hannah trained and practiced at levels never before seen. Hannah was the first true superstar of American supercross.

During the 1979 season, supercross grew explosively. The series had grown to 17 rounds, and the crowds were beginning to swell to massive proportions. On July 14, 1979, the largest supercross crowd of all-time was assembled when 74, 085 spectators were jammed into the Los Angeles Coliseum to watch the Superbowl of Supercross.

When Hannah was injured in a water skiing accident, the 1980 championship was open. Tall and lanky southern California hot-shot Mike Bell stepped up and took his factory Yamaha to the 1980 AMA Supercross Championship.

In 1981, Hannah made a gallant, much anticipated return to the sport. However, the veteran had a hard time getting himself, and his notoriously slow Yamaha, up to speed.

Additionally, Hannah was no match for new Suzuki whiz-kid Mark Barnett. Quiet and demure, Barnett was the exact opposite of the flamboyant Hannah. The Illinois-based rider was much more comfortable leaping into the night air towards victory than he was in front of the microphone. When all was said and done, Barnett claimed his first 250cc supercross title over Bell and Kent Howerton. Hannah ended up a distant fifth.

The 1982 title was won by yet another new, young rider, Donnie Hansen of Team Honda. Hot off of winning the 1981 Motocross and Trophy des Nations as a member of Team USA (along with Johnny O'Mara, Chuck Sun, and Danny LaPorte), Hansen kicked off the 1982 season with an impressive win before a standing-room only crowd of 71,000 fans at Anaheim Stadium. He would march on to win the title over Barnett and teammate O'Mara.

The era of 1983, 1984, and 1985 is, to this day, considered by most diehard supercross fans to be the greatest, most competitive, and most volatile era in the history of the sport. On any given night, any of 10 different racers could be considered likely to win. The list of great riders reads like a Who's Who of AMA professional supercross. Present in this amazing pool of talent were the ultra-smooth and stylish David Bailey, an older and more mature Bob Hannah (now on Team Honda), the quiet and consistent Barnett, the reserved and ubiquitous Bell, young hell-man rent-a-car destroying and style king Ron Lechien, the high-flying O'Mara, SoCal golden boy and proven winner Glover, the ultra-aggressive Rick Johnson of the El Cajon Zone (a notorious area of riding talent outside of San Diego), and the diminutive lightning-fast Jeff Ward.

In fact, with the exception of Bell, every one of the aforementioned riders would represent Team USA during its 13-year Motocross des Nations win streak. Known as the World Team Championship of Motocross, America would prove to the world every September that they were in fact the best motocross nation on earth.

Bailey would claim the 1982 AMA Supercross title and sadly have his career cut short a few years later after suffering paralyzing injures from a bad crash at a California winter series warm-up event in the tiny city of Huron. However, on a brighter note, the insightful and articulate Bailey later came back to the sport as a member of the ESPN supercross broadcast team.

O'Mara won the championship in 1984, while Ward won the 1985 campaign after a knock-down-drag-out season battle with Glover. The 1985 season culminated in a wild, one-race, winner-take-all battle at the season finale in the Pasadena Rose Bowl. O'Mara won the event, but Ward finished ahead of Glover and took home the hard-earned 1985 Supercross title.

The year 1985 would also be the year in which the AMA instituted the all-new 125cc East/West Supercross Series. Designed to develop young riders before they were to be thrown into the dog-eat-dog world of the 250cc ranks, the series was an immediate success, with Bob Moore winning the West Coast title (Moore would go on to win the FIM 125cc World Motocross Championship in 1994) and Eddie Warren clinching the East Coast Championship.

From that point forward, the 125cc series would prove to be an indelible element of the sport, with nearly every major modern-day American star, including the biggest of them all, Jeremy McGrath, doing time in the division.

Within this era, the Far East and the old world of Europe also became intrigued with the world of American supercross. International events sprang up all over the planet with the most formidable being the glamorous Paris Supercross and the world-renown Tokyo Supercross. The events are, to this day, hugely successful as the high-flying American racers are lured over to compete against the top European-based Grand Prix riders before thousands of adoring fans. Losses by American riders have been few and far between.

The 1986 season would mark the emergence of the next supercross superstar, Rick Johnson of Team Honda. Riding with flair, aggression, and calculated abandon, Johnson won 250cc Supercross titles in 1986 and 1988 before a broken wrist suffered during an outdoor national race at Gainesville stopped him from waltzing away with the 1989 title. Johnson's only serious competition during this time was Team Kawasaki's Jeff Ward, who defeated Johnson to win the 1987 Supercross title.

Johnson's wrist injury would eventually end his career, and Johnson's young teammate, Jeff Stanton, immediately picked up Team Honda's winning ways, clinching the 1989 title in commanding fashion. Stanton, a Michigan native, would back up his title in 1990 after a bitter, season-long battle with Honda teammate Jean-Michel Bayle of France.

A former World Champion, Bayle came to the United States to take on the fastest riders in the world. A win at the Gainesville outdoor national in 1989 on a privateer bike had proved that Bayle had the talent to compete in America. Riding for Team Honda in 1990, Bayle was in the hunt for the American 250cc title. Stanton prevailed, and the two riders became bitter rivals.

In 1991, Bayle ran away with the AMA Supercross Championship, proving that he was indeed the fastest rider on the planet. Bayle and his incredible talent changed the face of international motocross, and for the first time in a decade, put doubt into the minds of the

Jeff Ward took part in several of the most competitive seasons in supercross history. In the mid-1980s, any of a dozen racers ranging from young sensations like Rick Johnson to wily veterans such as Bob Hannah were candidates to win. *Paul Buckley*

Americans. In fact, Bayle set a precedent that resulted in a steady stream of European Grand Prix riders coming to race in the glamorous world of American supercross.

In 1992, Jeff Stanton would win the AMA Supercross Championship by a scant three points over the sensational Damon Bradshaw. After a season during which he won nine events, Bradshaw went into the season finale at the riot-battered Los Angeles Coliseum with a comfortable point lead over Stanton. A third-place position would clinch the title.

After rounding the first turn in a conservative fifth-place slot, it all went very wrong for Brad-

shaw. Visibly uncomfortable and shaken, Bradshaw began to unravel and make costly mistakes. Stanton raged along out front, followed by Mike Kiedrowski and Guy Cooper. Bradshaw never recovered and was eventually passed by a lackadaisical Bayle, who rode like his mind was on his upcoming Grand Prix roadracing career, and a young Jeff Emig, who was doing everything he could to stay behind his faltering Yamaha teammate. Stanton would go on to win the race and the title. Heartbroken, Bradshaw sat in the dark recesses of the box van in tears, while his mechanic, Brian Lunniss, threw wrenches around and cursed a blue streak.

Meanwhile, in the 125cc ranks, a young former BMX racer from Sun City, California, named Jeremy McGrath claimed his first Supercross Championship. McGrath dominated the West Coast series from the start, thus beginning a career that would soon boggle the minds of the entire motocross/supercross world.

The 1993 AMA Supercross Series opened at the Citrus Bowl in Orlando, Florida. The event was won by Mike LaRocco; the next weekend, Damon Bradshaw would win the Houston Supercross. Then, on January 23, came the Anaheim Supercross, and for all intent and purposes, the date marked a new era in American supercross.

After winning two titles in the 125cc Eastern Division, McGrath had been moved up to the premiere 250cc division for the 1993 season. In the third race of the 1993 season, McGrath showed up at Anaheim, his hometown supercross, relaxed and poised for battle. He won big. In fact, he would go on to win the next 9 of 13 events to completely dominate the series and claim the championship in his rookie year. From that point forward, McGrath was immediately cast into the lead role of American supercross. He would go on to dominate the 1994, 1995, and 1996 Supercross Championships, and along the way, also claim the 1994 and 1995 FIM-sanctioned World Supercross Championship (held during the AMA off-season).

However, the final and arguably the most-impressive act of the drama was yet to be played out. From the very first round of the 1996 season in Orlando, Florida, McGrath went on a rampage that resulted in 13 consecutive victories. Then, after finally being

After Ward won the 250cc Supercross title for Team Kawasaki in 1987, Team Honda took over and has won the 250cc Supercross title ever since. Jeff Stanton, shown above at Anaheim in 1993, won in 1989, 1990, and 1991.

Mike Kiedrowski winning his third consecutive Daytona Supercross in 1995.

Supercross has come a long way from its humble roots; now the Japanese manufacturers field multirider teams that can cost up to $4 million a season. Riders like Jeff Emig, shown here in 1994, can make a lucrative living as long as the appearances on the podium remain fairly steady.

beaten by arch-rival Jeff Emig in St. Louis, McGrath came back to win the season finale at Mile High Stadium in Denver. So it is now, with every AMA supercross record in his pocket, McGrath is considered the greatest supercross rider of all-time. To substantiate that claim, as if it needs to be proven, McGrath now owns the record for the most main event records in a season (14), most consecutive wins (13), most consecutive wins at one venue (7 at Anaheim), most supercross victories (43), and most AMA Supercross Championships (4). Truly amazing.

And now what has become of the sport of supercross some 25 years later? Perhaps our friend Ron Crandall sums it up best when he says, "Now I can watch it on TV, and I see it everywhere. There are events all around the world. I used to always make a joke in the rider's meetings and say 'Look guys, I'm getting kind of old here, and when I retire I would sure like to watch the sport on TV one night a week until I pass away,' and they would all kind of laugh, and now it is happening as we speak. I'm just loving it. We have taken something that was the brainchild of a couple of guys that were a little weird, but very, very bright, and now it has run the whole gamut. There are some great people involved in the industry, and the sport is just booming with TV, record crowds, 18-wheel trucks, and mainstream acceptance. I am very proud to have been a part of it all."

CHAPTER 2

HEROES

Supercross has changed dramatically since its conception in 1972, as have the riders who shaped the sport. Today's high-flying stars would be shocked by the low-budget factory teams and lower-key atmosphere of the early days of supercross racing. The riders from the early days, on the other hand, are just as amazed by the aerial acrobatics and 70-foot leaps performed routinely by modern supercross pilots.

What hasn't changed is that top riders are constantly redefining supercross racing, drawing the rest of the pack up to a higher level of riding. These riders become legends as their accomplishments are forever imbedded in the record books.

The following list are some of the racers who have made supercross what it is today. Some are high-profile celebrities—DeCoster, Johnson, and McGrath—who are worshipped on the shrines of podiums, trackside autograph sessions, and television cameras, while others contributions are lower profile but play a significant role in the history of the sport. Whether they are long gone and unknown to young fans or still making history, these riders are what supercross is all about.

Jean-Michel Bayle

Frenchman Jean-Michel Bayle, a 125 and 250cc World Champion, wasn't expected to be much of a threat when he came to the United States to race for American Honda in 1990. He was outstandingly successful, and JMB became one of the best the sport has ever seen. He cleaned up in 1991, winning the 250cc Supercross Championship, 250cc National Championship,

Damon Bradshaw making his comeback in 1995 at Mt. Morris. Bradshaw wore number 114 that year, and raised hopes that his talent and handlebar-dragging style could challenge Jeremy McGrath's dominance.

Bob Hannah was the first superstar of supercross, and dominated the series on Yamaha YZ250s in 1977, 1978, and 1979. In 1978, he won six consecutive supercross races. Hannah is shown on the podium at the Daytona Supercross later in his career. *Paul Buckley*

and 500cc Championship. In 1993, Bayle began a new career as a Grand Prix motorcycle roadracer, and has become one of the top 10 GP riders in the world.

David Bailey

The son of motocross instructor and track designer Gary "The Professor" Bailey, young David was one of the smoothest riders the sport ever produced. He won four Supercross and

National Championships before he was paralyzed attempting what witnesses described as an impossible double jump during the 1987 CMC Golden State Nationals. Bailey's accident did not end his role in supercross, however, as he went on to become a color commentator on supercross television broadcasts.

Mark Barnett

"The Bomber" Mark Barnett was the team captain when Suzuki was at its prime. From 1980 to 1982 he was unstoppable in the 125cc class, winning three consecutive titles. During that time, he also won 17 supercross main events, the 1981 championship, and was a member of the 1983 Trophy des Nations team. Now he has turned his talents to track building, where he assists super-cross promoters most of the season.

Mike Bell

Standing well over 6 feet tall, Mike Bell earned the nickname of "Too Tall." The lanky fac-tory Yamaha rider won the Supercross Champi-onship during what is considered one of the most competitive years ever, 1980. Throughout his career, he won a total of 11 250cc main events as a Team Yamaha rider. Now he works as a rep for Oakley, along with another supercross champ, Johnny O'Mara.

Damon Bradshaw

Hailing from Charlotte, North Carolina, Brad-shaw's on-the-edge riding style has made him a cult figure for race fans around the world. Brad-shaw has the dubious honor of being the rider with the most supercross wins in a season (a total of nine in 1992) without winning the champi-onship. Bradshaw retired a year later, saying he no longer enjoyed racing. In 1995, Bradshaw decided to come out of retirement and garnered solid fan support, but did not win a race.

Guy Cooper

Cooper is one of those riders whose story is much more than the sum of their win columns. While he never won a supercross main event, his 110- percent riding and stylish aerial antics made him one of the most popular riders in the sport. Cooper's only championship was the 1990 AMA

125cc National Championship. At the end of his motocross career, Cooper made the transition to off-road racing and won two ISDE gold medals along with several off-road races.

Roger DeCoster

In modern times, popularity is usually based on supercross wins. If that were true in the early days of the sport, DeCoster's one win during the 1974 Daytona Supercross wouldn't have gotten him very far. But DeCoster had one thing that the Jeremy McGraths and Bob Hannahs of the world didn't, he had five World Championship titles, and even won one of those after losing his spleen mid-season! DeCoster was later hired to be a race consultant for Team Honda and helped Team Manager Dave Arnold build the most powerful team in global history. After departing Honda, DeCoster was hired by Team Suzuki in 1995 to be team manager and helped develop a new line of RMs. The following year, he brought Suzuki its first 250cc National win in 15 years when his rider, Greg Albertyn, won at Unadilla in New York.

Tony DiStefano

Three consecutive 250cc National Championships gave Tony DiStefano a lot of respect in the motocross world. As the champion, he earned a spot on the 1975, 1976, and 1977 Trophy des Nations teams, one of which placed second. In the early 1980s, however, DiStefano was paralyzed while practicing in his backyard. But instead of leaving the sport behind, Tony D organized a plan to run motocross schools with the help of Suzuki.

John Dowd

Like Guy Cooper, Dowd got a late start in racing, and his success goes beyond championships. As the oldest factory rider on the circuit in the mid-1990s, Dowd's iron-man conditioning and late-moto surges made him a popular figure at tracks around the country.

Jeff Emig

A member of three winning Motocross des Nations teams, Emig has been one of the more dynamic riders of the 1990s. He won the 1992 AMA 125cc National Championship and narrowly beat Jeremy McGrath in the 1996 250cc outdoor

Johnny O'Mara won the 1984 250cc Supercross title on a Honda, a year after capturing the 125cc Outdoor National Championship. *Paul Buckley*

series. Emig was also the only rider to beat McGrath in the 1996 supercross season, and was one of the most consistent riders ever in the annual Motocross des Nations.

Broc Glover

The "Golden Boy," Broc Glover, will go down in history as one of Yamaha's most popular riders. As a teenager, he won three consecutive 125cc National championships (1978, 1979, and 1980) and then won the 1985 500cc National Champi-

Jeff Ward had one of the longest careers in supercross history and won each of the AMA's big four championships—the supercross and 125, 250, and 500cc outdoor national titles. He took home supercross titles in 1985 and in 1987.

onship riding an outdated air-cooled YZ490 in an era when Team Honda's David Bailey was riding a hand-built liquid-cooled works RC500 Honda. Glover also served his country proud by helping Team USA to the 1983 Trophy des Nations title.

Bob Hannah

Even with all of the hype surrounding Jeremy McGrath, Bob "Hurricane" Hannah is the icon of American motocross. In fact he single-handedly helped elevate salaries to the million-dollar state that has become commonplace since the early 1980s. To his credit, Hannah won 27 supercrosses, 3 supercross titles, 2 250cc National titles, and a 125cc National crown. He was even a member of three Motocross des Nations and Trophy des

Nations teams. Hannah's impressive performance in the Unadilla mud, aboard a 125 Suzuki, gave Team USA the win in 1987.

Kent Howerton

In 1976, Kent Howerton was on a roll in the 500cc class. He won the Outdoor National Championship and was on the Trophy des Nations team, which placed fifth. That was much better than the results of the 1975 Team USA squad, on which he served, which finished ninth. In 1980, Howerton won six of seven races to win the 250cc National Championship. He won five 250cc supercross main events during his long career (1979 Anaheim, 1980 Oakland, 1980 Pontiac, 1980 Kansas City, and 1981 Anaheim).

Rick Johnson won Supercross titles in 1986 and 1988 and was the first rider to break the records set by legend Bob Hannah. With a sunny disposition and a knack for self-promotion, Johnson quickly became one of the most popular supercross riders of all time. *Paul Buckley*

Jeff Stanton took over after Johnson's wrist injury took him out of contention, but the two riders were studies in contrast. Stanton was determined and consistent but not particularly flashy. Rough, difficult tracks favored Stanton's fanatical conditioning and mental discipline, and he won the Daytona Supercross four times from 1989 to 1992. *Paul Buckley*

Damon Huffman

Huffman put in two dominating years with Team Suzuki in the 125cc Western Region Supercross Series, winning the championship in 1994 and 1995. Smooth, precise, and stylish, Huffman ran and hid from the competition much as Jeremy McGrath was doing in the 250cc class. Should Huffman's talents translate to the larger bikes, Huffman is one of the young riders with the potential to be the next Jeremy McGrath.

Rick Johnson

Rick Johnson was adored by the fans, and he earned that honor with dominating wins and a sunny disposition. Johnson won five championships and perhaps would have won more if his career wasn't cut short due to injury. During a race in Gainesville, Florida, Johnson was accidentally hit by a privateer by the name of Danny Storbeck. The resultant wrist injury was too much to overcome, and RJ retired in 1992. Until Jeremy McGrath took over, Johnson's accomplishments made him the most celebrated supercross rider in American history. Johnson topped Bob Hannah's record of super-

cross wins by just one victory and set a new standard with 28 supercross victories.

Gary Jones

As a premiere 250cc rider, Gary Jones won the middleweight outdoor title during the first three years it was offered. After a stellar career, Jones became a test rider in the research and development department for White Brothers.

Pierre Karsmakers

Without doubt, Dutchman Pierre Karsmakers is one of the legends of supercross racing. He won the first-ever 250cc Supercross Championship in 1974, the first year that it was considered a true series. A year earlier, he had won the 1973 500cc Outdoor National Championship.

Mike Kiedrowski

The "MX Kied" dominated U.S. outdoor racing in the late 1980s and early 1990s. Despite his outdoor dominance, his success didn't transfer to supercross. A maniacal trainer and working man's hero, Kiedrowski retired at the end of a disappointing 1995 season. He decided to come back for the 1997 season and signed to ride for Honda of Troy.

Steve Lamson

Like Mike Kiedrowski, Lamson is a hard-working rider who has been dominant outdoors but ineffective indoors. His intense work ethic rewarded him with one of the greatest comebacks in motocross history, coming back from a knee injury and mechanical failure which cost him 75 points, to win the 1995 125cc National Championship. In 1996, he defended his title, wrapping up the championship with five motos remaining.

Mike LaRocco

Plagued by injury, LaRocco's determination has given him a reputation as one of the toughest riders in history. Mike's determination and ability

LEFT

Although Guy Cooper was not always the fastest guy on the track, he was consistently adored by his fans. Fearless in the air, Coop rarely came off a big jump without aerial acrobatics. After a successful career, Cooper switched from Suzuki's motocross/supercross team to the off-road squad.

to ride through the pain of injury has helped him win two championships aboard 250cc and 500cc Kawasakis. Mike's biggest disappointment came in 1992, during the 125cc National Championship points chase. He had a comfortable lead, but DNF'd three consecutive motos at the end of the year and lost a heartbreaker to Jeff Emig. In 1996, he switched to Team Suzuki.

Ron Lechien

Self-nicknamed "Dogger," Ron Lechien's natural talent rivals that of Jeremy McGrath. His tall, lanky body, combined with an ultra-smooth riding style, made him one of the most exciting supercross riders ever. He won his first supercross at the tender age of 16 and went on to win the 1985 AMA 125cc National Championship. Four years later, he was the top American at the Motocross des Nations and, in fact, was the fastest rider of the event, period. Unfortunately, a broken femur at Steel City in Pennsylvania ended his career at the turn of the decade.

Jeremy McGrath

McGrath is the defining star of supercross. With an unprecedented four Supercross titles under his belt, he has shattered the records set by legends Hannah and Johnson and become the most dominant supercross rider of all time.

A converted BMX racer, McGrath first made his mark when he won two 125cc Western Region Supercross Championships. McGrath's shortcomings as an outdoor rider—he consistently finished outside the top five—downplayed his unquestionable speed indoors. Still, when Jean-Michel Bayle left Team Honda in 1993, McGrath was brought up to the premier class by team manager Dave Arnold, who saw the potential in the kid from Sun City, California.

To everyone's astonishment, McGrath won the championship during his rookie year (and managed to squeeze in a single 125cc National win). McGrath's reign as the king of the stadiums continued through four seasons of mastery, capped off by a stunning 1996 season that saw him win 14 of 15 rounds.

"Showtime" completed his mastery of the sport in 1995, winning the 250cc Outdoor National Title with a hard-charging style. Despite losing the 1996 outdoor title due to an injury and

A natural, consistent rider and charismatic extrovert, Jeremy McGrath has surpassed all the supercross records. McGrath dominated the sport since his rookie title in 1993, prompting him to wear motocross pants bearing, "RJ Who?" for the race in which he surpassed Rick Johnson's supercross victory record.

Jeff Emig's determined, consistent riding, McGrath was clearly the fastest man on the track. In the 1996 Motocross des Nations, McGrath proved he was indeed the fastest rider on the planet, easily besting the top riders from around the world.

Johnny O'Mara

Known for his smooth and flashy style, O'Mara was a member of Team Honda during the boom years. O'Mara won the 1983 AMA 125cc National Championship and a year later captured

the 250cc Supercross title. He retired in 1992 after a short stint with Team Kawasaki, but then came out of retirement in 1994 to ride for Honda of Troy for one year.

Mickael Pichon

Following in the footsteps of fellow French countrymen Jean-Michel Bayle, Pichon came to the United States in 1994 and won the 125cc class at the San Diego Supercross on a privateer Honda. Then he disappeared for a year to race the GP circuit in Europe. After limited success, he was hired by Team SplitFire and won the 125cc Eastern Region Supercross Championship in 1995 and 1996. His two titles eventually landed him a spot on Roger DeCoster's Team Suzuki in 1997 as a 250cc rider.

Donny Schmit

Two World Championship titles made Schmit one of the most successful American GP racers of all-time. After winning his world championships, Schmit returned to the States racing four-strokes and enjoying life in his home state of Minnesota (including constructing snow supercross courses). In 1996, a few days before the Minneapolis Supercross, Schmit died suddenly due to a rare disease.

Gary Semics

Hailed as the sport's first-ever 500cc Supercross Champion, Semics won the inaugural title in a tie-breaker. He had tied Tim Hart in championship points, but was crowned champion because he had posted the most top-three finishes.

Marty Smith

In 1974 and 1975, Smith won a pair of AMA 125cc National Championships during the first two seasons of the series. A year later, Smith faced off with Bob Hannah, and Smith's dominance ended when Hannah beat him 347 to 260. Smith still finished second overall. In 1977, he fought back and captured the 500cc National Championship to collect his final of three titles. All told, Smith was the first big superstar and matinee idol of American motocross.

Jeff Stanton

Michigan's Stanton was one of the hardest-working riders the sport has ever seen. Taken under fellow Team Honda member Rick Johnson's wing in the late 1980s, Stanton quickly matured into a steady, determined, but not particularly stylish rider. He won six championships during his career. After retiring at the end of the 1994 sea-son, Jeff signed on as a consultant for Team Honda, and serves as a sideline coach during supercross and outdoor races.

Marty Tripes

Tripes, a factory Honda rider, is credited with winning the first ever supercross on July 17, 1973. In actuality it was the third such race, but the first two events were held at Daytona International Speedway, and because of the speedway's enormous size, it wasn't considered to be a true supercross. Tripes' win came inside the legendary Los Angeles Coliseum, and he topped Jim Pomeroy, Antonio Baborovsky, Jaroslav Falta, and Bob Grossi.

Jeff Ward

Known as the "Flying Freckle," Jeff Ward's short stature and fluid riding gave him a unique style that produced seven National and Supercross Championships. Wardy is the only rider to have won all three Outdoor Championships (125, 250, and 500cc) plus a Supercross title. He was also on seven winning Trophy des Nations and Motocross des Nations, a feat that no other rider can boast.

Jimmy Wienert

Jammin' Jimmy Wienert was at his prime in the late 1970s. He was the third-ever Supercross Champion and two years earlier he had scored a pair of championships in the 500cc outdoor series. Because he was such a force, he was chosen to represent Team USA during the Trophy des Nations. Teamed with Jim Pomeroy, Brad Lackey, and Tony DiStefano, the team placed second.

Kevin Windham

In 1996, Windham won the 125cc Western Region Supercross Championship and won several outdoor nationals (and was the only 125cc rider with the speed to challenge Lamson); pretty impressive considering that he was only 18 years old. The "Ragin' Cajun" is another likely candidate to take over the helm when Jeremy McGrath retires.

CHAPTER 3

FACTORY BIKES

When Jeremy McGrath's Honda CR250R is rolled to the line before a race, it is immaculately prepared. His full-time Honda mechanic, Skip Norfolk, has spent hours pouring over every bearing, bushing, nut, and bolt to ensure that it makes it across the finish line without a single hiccup. Between motos, the bike is fitted with new tires, the clutch plates are inspected, and any items with questionable mechanical integrity are instantly replaced. Between races, the motorcycles are stripped and every critical part is carefully inspected and replaced at the first sign of wear. Every time McGrath comes to the line, he's essentially racing a brand-new bike.

McGrath's race machine does not look much different from the CR250 available at the Honda dealer, and in many ways, it's not. Unlike almost every other form of motor racing, supercross and motocross racers use relatively stock machinery. The infamous production rule (more on that later) of 1986, dictates that factory racers must use relatively stock machinery.

One other way that supercross and motocross are unique is that the difference in machinery is relatively minor. The rider is the most important difference, with technology playing a more subtle role. Although Honda has always had some of the best bikes out there, it is generally acknowledged that Jeremy McGrath would win on any of the bikes on the line at a supercross race.

At the 1996 Paris Supercross, McGrath and Steve Lamson ended up on Honda of Troy bikes due to some difficulties with the Honda factory. Neither rider was slowed down. Factory

Factory race bikes have always been and will always be the pinnacle of performance, but the radical works bikes of the mid-1980s prompted the AMA to require manufacturers to compete on production-based motorcycles. *Paul Buckley*

equipment is an advantage, but the benefits of full-time, top-notch mechanics; advice from veteran managers; full-time trainers; and plush travel accommodations have a greater impact come race day.

That said, factory race bikes are still the trickest machinery on the track. The maxim, "win on Sunday, sell on Monday," holds true, and factory teams come to the track to win. With big budgets and championships in mind, the factory bikes are loaded with handmade, one-of-a-kind parts.

While all the factory bikes are full of high-tech bits, the Hondas have a special appeal, due to the combination of a tradition of winning championships and a reputation of spending vast amounts of money on top-notch equipment.

"Works" means parts rather than exotic chassis designs and one-off engines, but the parts themselves are nothing short of state-of-the-art. They are the compilation of two decades of testing in Honda's extensive research and development program. A program run by some of the best motorcycle engineers on the planet who have been hand-selected with one goal: to build the best bikes.

McGrath's 1996 CR250 was, of course, fitted with an extremely healthy motor. Honda of Japan ports the cylinder and modifies the head. The piston and connecting rod were factory pieces, and the pipe, airbox, and carburetor were also Honda of Japan parts, with a carbon-fiber silencer shaving a few ounces of weight. The power valve was modified, and the Powerjet carb that appeared on the 1997 CR250 was on factory Honda bikes at least a year before they appeared in production. The ignition was also extremely trick, and timing can be altered with a laptop computer. Gear ratios were juggled internally, and factory covers give the engine that unobtainable look.

The chassis features works Showa suspension, often cited as the largest advantage factory riders have over the privateers. The brakes are works items, and the bike is lightened with numerous titanium and aluminum fasteners as well as a sprinkling of carbon-fiber pieces.

The other factory bikes use a similar array of modifications, with engines and suspension heavily modified internally and trick brakes, carburetors, exhaust, and fasteners finishing off the bike.

Some bikes feature other extremely trick bits. The 1996 Kawasaki SR250 used a traction-control device that altered the timing curve. Using a handlebar-mounted switch, the timing curve could be altered to apply smoother power to the rear wheel. The swingarm on that bike was lengthened 10 millimeters. This chassis modification would be considered illegal, but Kawasaki offered the same part for sale to avoid problems. The long swingarm would set you back $850 at the dealership.

The Yamaha 250s of the mid-1990s sported the famed long-rod kit, which was the hot setup for holeshots in the 1995 race season. The kit was available from the dealership and showed up in more than a few local racers' bikes.

In addition, factory riders use any number of aftermarket parts on their bikes. Part of this is due to sponsorship dollars, and part due to increased reliability and longevity.

Early Works Bikes

The trick bikes of today are a far cry from the early years of motocross. The bikes were comparatively crude, and the generous budgets and extensive research and development of today were nonexistent. Most riders served as their own mechanics and went back to a regular job on Monday. Riders often raced the same motorcycle throughout the entire season and had to choose between a rebuilt top-end and supper for the next month.

In the early years, the fastest bikes were four-strokes. Four-strokes became less common during the late 1960s and were outclassed by the lighter, more responsive two-stroke racers by the 1970s. Tractable power and reliability kept four-strokes a favored mount for off-road and play riders, but thumpers have not been competitive in motocross for almost two decades. Recently, however, four-strokes have made a comeback. Belgian rider Joel Smets won the 1995 500cc World Championship on a 501cc Husaberg and then won the Open class at the Motocross des Nations in the same season.

In the late 1960s, though, four-strokes were king until the Japanese manufacturers—most notably Suzuki—seriously entered the racing arena. The level of technology and reliability

Mike Bell's works Yamaha was one of the trickest bikes on the track. It featured a hand-formed aluminum gas tank, magnesium cases and hubs, and a hand-built one-off factory swingarm. Yamaha pioneered widespread production use of water-cooled engines.

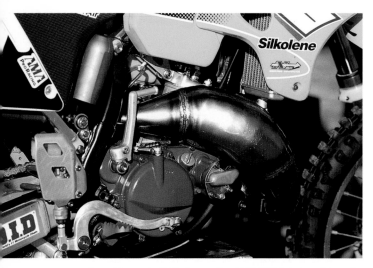

Greg Albertyn's 1996 factory Suzuki RM250 bristles with works parts such as the high-tech side case, titanium footpegs, and factory-spec pipe. Like all true works parts, these are not available to the general public.

quickly took an upward turn, and the works bike emerged. Works bikes were machines constructed of one-off hand-built parts that were unobtainable by the general public. The object was to offer "factory" riders a distinct advantage over the competition.

In the early years of racing, one of the primary concerns was to reduce the overall weight of the bikes. Many of the production bikes were built like Sherman tanks. Reducing the weight 20 pounds or so changed the way they could be ridden and reduced stress on other parts. Weight was shaved with featherweight magnesium and titanium parts. Mechanics also trimmed excess material off of existing parts.

Throughout the 1970s, the weight of works and production bikes continued to drop, and the level of works bikes gradually rose. The most significant development of the 1970s was the rise of long-travel suspension. The manufacturers learned the importance of quality suspension when Hakan Andersen won the 1974 World Championship on the first factory Yamaha with single-shock suspension. Greater suspension travel meant faster lap times, and state-of-the-art changed from the Husky and CZ

with 4 inches of travel to 12-13 inches of travel by the early 1980s.

The Works Era

Beginning in the late 1970s, the level of the factory bike rose to an unprecedented level. Manufacturers began building one-off machines that gave their riders insurmountable advantages on race day. In the early 1980s, factory bike development went through the roof and the modern dirt bike emerged, but it was only in the hands of those with factory rides. Single-shock rear suspensions, water-cooled engines, power valves, disc brakes, and relocated fuel tanks are all features that were developed in this era.

Rear suspension was radically changed when the single-shock systems appeared. Yamaha developed the Monoshock setup in the 1970s, but it raised the center of gravity and did not offer a progressive rate. By mounting a single, short shock vertically with a variable-rate linkage, the weight was moved low and the suspension stayed supple on stutter bumps while soaking up the big hits as well. A remote reservoir kept the oil and gas cooler, reducing shock fade. Suzuki had a unique twin-strut system called the Full-Floater, and Yamaha had a rather odd-looking link system which it introduced in 1982. Kawasaki's Uni-Trak was another early single-shock arrangement. These systems pioneered external compression and rebound adjusters that allowed the suspension to be tailored for each specific track.

In 1983, Suzuki built Mark Barnett a bike on which the locations of the fuel tank and air filter had been swapped. The object was to lower the center of gravity, but this in itself created another problem. Suzuki had to use a fuel pump to deliver gas to the carburetor, which is normally gravity-fed.

Once weight was down, the manufacturers went to work on the engine and concentrated on port design, carburetion, exhaust systems, internal gear-ratio changes, and water-cooling (instead of air-cooling). At one point in the early 1980s, Honda actually began experimenting with a rudimentary form of motor and suspension telemetry, 'a la Indy cars of today.

In 1982, power valves began to make their way into production as a direct result of racing. Yamaha created the YPVS (Yamaha Power Valve

This is the most dominant bike in supercross history: Jeremy McGrath's 1996 CR250R. Tuned by Skip Norfolk, it was one of the first bikes that successfully used a Powerjet carburetor. Both Suzuki and Kawasaki tested similar units on the race track but encountered teething problems and were forced to remove the system for further development testing. The Powerjet carburetor appeared on the 1997 production CR250s.

Sights such as this drawer full of titanium bolts are commonplace inside Kawasaki's race shop in Irvine, California. If you have to ask how much the ti bolts cost, you can't afford them. The seat for Jeff Emig's 1996 Motocross des Nations KX500 rests on top of the bin.

Mechanic Ian Harrison, an accomplished motocross rider, works on fellow South African Greg Albertyn's factory RM250. After an error-filled 1996 season, Albertyn opened the 1997 season with a supercross win at the Los Angeles Coliseum.

System), and two years later Honda introduced the ATAC (Automatic Torque Amplification Chamber). The idea behind these variable exhaust port height valves was to adjust the size of the exhaust port as engine rpm changes. A large exhaust port yields better high-rpm horsepower, while a small exhaust port gives the engine more low-end power. The power valves gave the best of both worlds, added much-needed torque and low-end boost to the traditionally peaky and difficult-to-ride two-stroke racing engines.

Most of the factory race bikes had handmade aluminum fuel tanks, works suspension components, and some had early front and rear disc brakes. Honda was so dedicated to its race program that its racing version of the quarter-liter CR, dubbed the NT3 RC250, was night and day different cosmetically and mechanically from the version that consumers could buy at the dealers.

Yamaha's bikes looked the most like their production versions but were tricked out with high-tech gadgets. Depending on course conditions, the engineers had developed two different swingarms to change the wheelbase of the bike for supercross and outdoor-style courses. They also utilized 40 millimeter Mikuni carburetors (instead of the stock 36 millimeter unit). Perhaps the trickest Yamaha was Ron Lechien's works YZ125, which featured rotary valve induction instead of reed valves.

In 1985, Honda introduced an electronic ATAC power valve system. It required a NiCad battery (mounted near the carburetor and concealed in an aluminum box) and a series of relays that led to a handlebar-mounted on/off switch. They also tossed the stock radiators and hand-built an oversized model for the right side, almost 18 inches long, and shortened the left side to almost 10 inches. It was odd but allowed Honda to lower the center of gravity and make a much larger expansion chamber on the exhaust system.

RIGHT
Austrian-based KTM's factory bikes sport works forks, machined billet triple clamps, hand-made exhaust pipes, magnesium outer cases, and an Ohlins shock. Although KTMs dominate the off-road and European scene, but have a low profile in American motocross.

This high-tech scrap pile contains over $7,000 worth of fork tubes that have been used by Team Kawasaki riders damaged and beyond repair.

These Kawasaki swingarms were used during preseason testing. The result was the late release of a 10mm-longer swingarm for the 1996 season.

This also required construction of an all-new aluminum tank and a safety-type seat that ran all the way to the steering stem. Twin-piston disc brakes front and rear were in place and Honda engineers constructed a fiberglass airbox.

The rising level of technology drew the attention of motocross and supercross racing's sanctioning organization, the American Motorcyclist Association (AMA). Privateers were at an increasing disadvantage due to the ultra-trick factory racers, and the AMA looked for ways to level the playing field.

One system was the claiming rule, in which privateers could purchase works bikes for a set price under certain conditions. This did not prove to be

enough of a deterrent for the AMA, and, in 1986, Big Brother stepped in. Concerned about the future of racing, the AMA instigated the production rule. With the exception of items built for safety reasons, the rule severely limited the amount of exotic works parts that could be used. In essence the rule was designed to level the playing field so that privateers would have a chance. How did it work? It depends on how you look at it. Factory riders were still winning, but the reason had more to do with hiring talented riders than riding bikes that offered an unfair advantage. The production rule did, however, force the manufacturers to build better production bikes. That meant that companies like Honda, Yamaha, Kawasaki, and

Jeff Emig's Jeremy Albrecht-tuned 1996 Kawasaki SR250 was the only bike other than Jeremy McGrath's CR250 to win a supercross all season long. For the record, the SR is Kawasaki's designation for its works KX models, and it stands for "Special Racer." Kawasaki is the only manufacturer to ship its race bikes from Japan, built to works specs. The other manufacturers begin with production bikes stateside and then add works components.

Suzuki put their secrets into production sooner so that their riders had the best bikes on the track.

The decade following 1986 saw a lot of refinements and miscellaneous changes, especially in the area of suspension. The fork was inverted, and then turned right side up again. The past decade has also seen circulating and non-circulating systems, several cartridge designs, and a twin-chamber system. Changes to the rear suspension were less visible, but it too was dramatically improved through radical linkage changes and external adjustments. As of 1996, Honda

PARTS GALORE

Factory race teams spend millions of dollars in the quest for a championship. The last thing they want is to lose a race due to a mechanical failure. As a result, they change hard parts, like frames, with the same frequency a C rider changes tires.

The bikes are torn apart after every race, and anything that looks worn, cracked, or fatigued is replaced. Though it varies a bit among teams, this is the frequency that most parts are changed even when damage is not detectable:

Part	Frequency Changed
Frame	4-6 races
Handlebars	4-6 races
Foot pegs	10 races
Triple clamps	30 races
Swingarm	20-30 races
Wheels	20-30 races
Tires	Every moto
Cylinders	4-6 races
Power valves	4-6 races
Piston and rings	Every week
Clutch	Every moto
Gear oil	Every moto
Spark plugs	Every week
Air filter	Every moto
Levers	4-6 races
Clutch cable	Every week
Grips	Every week
Brake pads	Every week
Chains	Every week
Sprockets	Every week
Graphics	Every week
Plastic	Every week
Suspension fluid	Every week

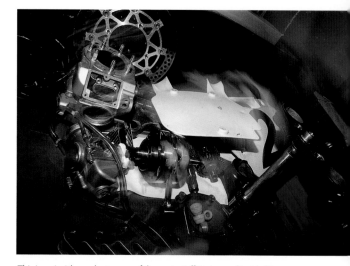

This is an inside peek at many of the parts Jeff Emig used inside his race bike during the 1996 season. Everything from brake rotor size to transmission gear ratios have been altered, often with hand-built parts.

Second-level teams like Honda of Troy enjoy limited factory support and sign lesser-known up-and-coming riders or, more typically, ex-factory riders that want to lengthen their careers. The combination makes many of these smaller teams capable of regular appearances on the podium. Honda of Troy is the highest-profile nonfactory team, and ex-factory stars Larry Ward, Mike Craig, and Mike Kiedrowski have been part of the HoT squad.

Honda revolutionized the industry in 1997 by introducing a new CR250 equipped with a box aluminum chassis and Powerjet carburetor. Here, Jeremy McGrath airs out the new bike at the 1996 Paris Supercross. This is one of few times McGrath appeared on a 1997 Honda; he switched to Suzukis for 1997.

offered high- and low-speed compression adjustments on the rear Kayaba.

Chassis also underwent change. Kawasaki created the perimeter frame in the early 1990s, and during the same period nearly every motocrosser came standard with a removable subframe to make air filter maintenance and shock changes much easier. Then in 1997 Honda dug deep when it decided to build its highly innovative CR250. Years of race testing in Japan by Suzuki, Yamaha, and Honda had proven that they could build a road racing-inspired twin-spar aluminum-frame motocross bike, and they did. They combined this with two other production firsts: a

Powerjet carburetor and an electronic traction-control device.

So what does the future hold? It kind of depends on legislation. There is a big push in California to clean up the air, and there has already been legislation written that would kill two-stroke engines. Honda has also been experimenting with cleaner-burning motors, but the jury is still out as to whether or not it will be enough. Even if two-strokes do become extinct, technology will march on with four-strokes. Many industry experts feel that the next significant changes will come in suspension, but no one knows for sure . . . as of now, that is a trade secret.

CHAPTER 4

RACE WEEK

From the spectator's point of view, professional racers go to work Saturday night when the gate drops and are done at the checkered flag of the final moto. Although it's true that races are what a pro rider's week is all about, successful racers are on the job seven days a week.

Twenty years ago, riders and their mechanics (if they were lucky enough to have one) spent the better part of the week on the road, traveling from race to race. Nowadays, however, some factory mechanics and most of the riders fly to the races each week, which allows them to utilize their time more efficiently. This is one of the bigger advantages that the works riders have over their privateer counterparts.

Typically, on Sunday the riders and mechanics will fly home from the races, and when you factor in plane changes, layovers, and other delays, this alone can consume an entire day. At best they will have just enough time to do laundry, grab some dinner, and take a brief mental break.

On Monday the work begins. At the beginning of the season, the time is usually spent at the team's private supercross track to test new parts and to work on technique. This also allows the riders to practice on a real supercross track, which is a huge advantage that the larger race teams have in their favor. The teams will repeat this process on Tuesday and Wednesday, and the riders will usually do some type of cross training (either bicycling, running, or gym work) in the morning or late afternoon to keep themselves in top-notch condition.

At the beginning of the season, most teams spend four solid months testing equipment for the upcoming year. Team Suzuki's Mike LaRocco is one of the most dedicated racers and uses the precious time to his utmost advantage, as this photo, taken at Suzuki's private test facility in San Bernardino, California, shows.

During the week, riders like Jeff Emig usually practice three times, and that leaves little time to maintain equipment. Usually Emig has only one bike at his SoCal home, but in this case he also had a KX500 for his assault during the 1996 Motocross des Nations in Jerez, Spain, where he won the Open class and helped bring the team championship back to the United States.

On Wednesday, one rider from each team usually flies out with all of the mechanics to the race site. While the mechanics prepare the bikes, the chosen rider will participate in a press conference for local newspapers and television stations on Thursday. The chosen riders of the teams usually have an opportunity to ride the track for the very first time, which is usually in extremely crude condition. This allows members of the press an opportunity to take photographs and/or video tape riders, which is used to preview the race and possibly increase ticket sales. Riders rotate press-day duty, among each team, on a weekly basis.

The team managers and the rest of the riders usually travel to the race on Thursday. This will give them just enough time to prepare for Friday's practice day, which is the first time that the bulk of the riders have an opportunity to ride the course.

Come Friday night, the mechanics go over their bikes and make sure they are race-ready. Though they will have a little time to work on the bikes on Saturday morning, Friday is usually the last opportunity for any major change. After the race, this hectic schedule repeats itself for nearly 30 weeks, with only three or four breaks. It's an intense schedule, but it's all part of being a professional racer.

Building Tracks

In most cases, it takes an entire week for workers to build each supercross track, and they

Between races, factory mechanics fly home to work on new machinery that will be used for testing, photo shoots, or special races. The actual race equipment stays on the road inside the team's $350,000 transporter. Interestingly enough, the riders don't actually get a whole bunch of new bikes throughout the season. Instead, parts are changed whenever necessary to ensure that everything is in top-notch working order.

Wyatt Seals tore Ryan Hughes' bike all the way down after every race. Here, he buffs the swingarm with a wire brush for that way-cool factory look. Seals was recruited to wrench for McGrath in 1997.

Kawasaki scratch and dent case displays parts destroyed in testing.

This is the second story of Team Kawaki's high-tech transporter.

are always under the gun due to time constraints, especially when it requires covering up grass. First, the area is blanketed with 4x8-foot sections of plywood, and then a layer of plastic is laid down to make the cleanup process easier. The first two steps alone can take a full day. If the stadium is open to the ravages of Mother Nature, the weather can easily slow down this process!

The most time-consuming part of the process is bringing in the dirt, which requires a hundred or so truckloads of soil, which is either borrowed or rented. Obtaining good dirt in this quantity can be so difficult that the promoters often times will lease space just so they can use the same dirt every year. Sure, it's expensive, but it is one of the only ways to guarantee a good surface.

Once the dirt is brought into the stadium, it is scattered around by dump trucks to form a base that's approximately 6 to 8 inches deep. It has to be thick enough so that when ruts develop, the riders won't dig into the plastic. After the base is formed, the track is laid out according to blueprints and this is done using lime (just like in baseball) and stakes. After this stage, the final dirt is imported, and precise quantities are dumped according to size and type of obstacle.

From the various piles, a half-dozen skid-loader operators move it around and form the actual obstacles according to the track diagram. This is the most difficult part of the process, since very few heavy-equipment operators possess the knowledge to build obstacles according to supercross specs. That's why the promoters hire the same crews to construct the tracks in each city. This will save a lot of time and avoid last-minute changes. And if there are any alterations that need to be made, they are usually discovered and fixed during Friday's practice session. The AMA doesn't like to make any major modifications on race day unless the changes are for safety reasons.

Unfortunately, things don't always go according to plan. During years when the Minneapolis Supercross was held in February, the snow-covered soil was frozen, and it was nearly impossible to create obstacles with it. A similar problem happened at the 1995 Anaheim Supercross when heavy rains turned the track into soup and track builders struggled

Team SplitFire's brightly colored rig attracts a lot of attention and serves as a workshop on wheels at the races. The team has had some of the fastest, best-looking, and most popular bikes and riders on the track. Ryan Hughes and the Team SplitFire/Hot Wheels KX125 was one, as was 125cc Eastern Region Supercross Champion Mickael Pichon and his chrome-plated KX125.

to form any obstacles that would survive practice, let alone racing.

Unless there is an amateur event on Sunday, the crew usually begins dismantling the track a half-hour after the races are over. They only have a short time to get the stadium cleaned up before they have to fly to the next city and begin constructing an entirely different circuit.

The Folks Behind the Stars

From your seat inside the stadium at any supercross, the life of a factory rider appears almost make-believe, but what you don't see during the three hours is simply astonishing.

There is another entire world to which most people are oblivious, except for die-hard race fans. The world is the pro pits, and access is strictly limited to those whose lives and jobs revolve around motocross and supercross racing.

On race days, from sunrise to well after midnight, the track becomes a magical place full of intrigue; after all, the place is surrounded by secrecy as teams try to maintain even the slightest technological edge over their competition.

But who are all of those people inside the no-man's land, and what do they do? Why is there usually 750 credentialed personnel in the pits when there's only 80 spots in the program? It

Skip Norfolk (left) and Mike Gosslaar examine a works clutch. Mechanics often have to work together in order to solve problems fast and effectively.

Even with a massive transporter, the pits can become a very crowded place during business hours.

isn't like NASCAR or Indy car racing where each team requires a minimum of a dozen people to work on the racing machine. But it does require a larger group of people than one might expect.

Each factory race team has a manager, a mechanic assigned to each rider, a motor man, a suspension technician, and a driver if the team is fortunate enough to own a semi. It's a tightly knit group that depends on one another for continued success. Like most motorsports racing, supercross is essentially a team sport, although individuals receive the bulk of the glory.

One of the keys to a successful effort is the information that each rider passes along to their mechanics and engineers. Even though each rider's fate lies in their own hands once the starting gate drops, every bit of information learned is another valuable piece of the puzzle. Help in any area can mean the difference between winning and losing, especially when it comes to the never-ending search for the perfect suspension setup. Therefore, communication is the groundwork for every winning race effort, whether it is two- or four-wheel racing, and it requires a lot of people to wade through the data and determine areas that need improvement.

The communication extends well beyond the team boundaries of Yamaha, Honda, Kawasaki, Suzuki, and KTM. At each event, there are representatives from tire manufacturers and other

After the races, it usually takes about two hours to tear down the awning and put away all of the race equipment. These Yamaha folks are packing up in the rain that followed the 1996 Anaheim Supercross.

aftermarket companies. It's up to these people to supply the race teams and privateers with hard parts such as handlebars, sprockets, chains, plastic, graphics, pistons, brake pads, and so on.

Then there's a whole separate army that caters specifically to the riders and their apparel needs. Each weekend the goggle manufacturers hand out nearly 150 hand-prepped lint-free goggles fully loaded with tear-offs; gear manufacturers deliver custom-made pants and jerseys; and each helmet manufacturer has a representative whose sole responsibility at the

Inside all factory transporters there is a lounge that serves as a retreat for riders and team personnel. This one was designed by Kawasaki Team Manager Roy Turner and has all the comforts of home, including a microwave oven, satellite television, VCR, bathroom, shower, and refrigerator.

Bike setup is very critical, and nothing can be left to chance. Interaction between riders and mechanics, and Jeff Emig like Jeremy Albrecht, is important to a winning race effort.

LEFT
Though it may not appear this way on the track, most of the riders are actually good friends and spend a lot of time together. In fact, Jimmy Button occasionally stays over at McGrath's half-million-dollar home when he's racing in California.

races is to clean and maintain helmets for its team riders.

Aside from supplying riders and race teams with products, the job of a manufacturer representative is more than handing out a bunch of freebies to a hard-core group of heroes. These same people also have to recruit up-and-coming talent. It's a job that requires tremendous insight and the ability to forecast who has the ability to be champions in the future. It is a task more difficult than predicting the weather, and oftentimes it can be a cutthroat business. Timing can be everything, and it has become such a good business for riders that nearly all of the top pilots have hired business managers to take care of their finances. It's this type of interaction that typically makes it mandatory to keep the pits closed. After all, business is business, and the pro pits are where some of the most important business decisions are made.

CHAPTER 5

RACE DAY

For the average spectator, race day begins in the late afternoon after paying seven dollars for parking on the stadium grounds. The first order of business is to prepare for a tailgate party or to figure out how to gain access into the pits for a behind-the-scenes look at the sport.

But for race participants, the day begins at the crack of dawn when alarm clocks wake up race team members, who then begin a weekly ritual. Mechanics, technicians, and crew members stumble out of their hotel rooms around 6 A.M. If they are lucky, they will hit a hotel restaurant or a nearby coffee shop for breakfast, but most likely they will head straight to the stadium to begin setting up the pits. In a matter of hours, the stadium parking lot is transformed into a small community of the sport's elite. A half-dozen semis, 40

or so box vans, and a few pickup trucks and vans set up camp, which will only be there 18 hours. The pits are as temporary as the man-made track, and play a pivotal role in the events that lie ahead.

Around 8 A.M., the AMA opens tech inspection, and each bike is paraded through an assembly line of eagle-eyed inspectors whose sole job is to detect any cheating and insure rider safety. The bikes are weighed to ensure they meet the minimum weight requirement, and then they are tagged for identification purposes.

The riders don't actually start trickling in until about 10 A.M. Since most of them have pre-entered, they head straight toward the pits to check in with their teams to make sure there aren't any problems or changes in the day's activities. It's also during the pre-noon hours that most

Jeremy McGrath, the former BMX racer, is so dominant that during the 1996 season he only lost one race (he finished second at St. Louis). It's doubtful that many of his records will ever be broken.

RIVERDALE PUBLIC LIBRARY DISTRICT

49

Supercross brings motocross racing stars to the masses, filling stadiums across the country. This 70,000-seat palace is Anaheim Stadium.

Damon Bradshaw spent most of his life racing motocross, and won his first 250cc Supercross at age 17. Bradshaw suffered a huge setback in 1992 when he lost the Supercross Championship after winning what was then a record nine races in a single season. Shortly thereafter, Bradshaw retired as a millionaire at age 22. He returned to racing a year and a half later, but at the conclusion of the 1996 season, Bradshaw parted company with longtime sponsor Yamaha after a disappointing comeback.

riders begin the important ritual of walking the track. Savvy racers spend well over an hour or so looking for hot lines that could later prove to be the difference between winning and losing. The most serious riders even go up into the stands to get a bird's-eye view of the track layout, since some of the more obvious lines aren't visible from ground level.

Most teams hire former racers to offer advice to the younger generation of hotshots. In 1996, Yamaha had the legendary Bob Hannah, Honda had Jeff Stanton, Suzuki had Roger DeCoster, and Kawasaki had Mike Kiedrowski. Their role is very important in the day's strategy. Prior to most races, the riders are only given an hour to practice on the race track, and that is broken up into several sessions on press day, a special practice day, and race day. During this time the track will undergo several changes as the track's construction crew works to perfect the racecourse design. Oftentimes, complete sections have to be reshaped to improve lines that will make for better race action. It's up to the riders and their teams to keep a close eye on the changes to see how it will affect their approach to the final and very brief practice session. This is the last time that riders will have to attempt some of the larger double and triple jumps before race time.

Moments before practice, some of the final adjustments are made to the bike. This is the time when shock sag is set, the controls (handlebar, clutch, and brake levers) adjusted, and so on. From here, the riders and mechanics wait in line for their shot at the first of two race-day practice sessions.

To most riders, practice is fairly routine. The goal is to learn the course as quickly as possible and try to charge from the beginning. On the other hand, the elite have a totally different approach and conform to a strategy that their peers have developed over the years. They use the first session to look for good lines and to make adjustments to suspension, chassis, and engine setup. All the while, the other members of their teams are videotaping a sampling of fast riders and, in general, are observing what works and what doesn't.

It's only during the final practice session that many of the fast guys bust out lines which they intend to use during the race. Oftentimes this includes spectacular combination jumps that are turned into unreal double or triple jumps. The strategy is to wait long enough, so that the competition doesn't have enough time to try out the lines and then are faced with trying them in their heat races or semi races if ever at all. In essence, it's a poker trick; don't show your hand until absolutely necessary.

After the two practice sessions, the final preparations for the evening are made. While the riders work with sponsors to get their goggles prepped, helmets cleaned, and riding gear checked, their mechanics wash the bikes and then tear them down to ensure mechanical integrity. During this time, new tires are installed, and clutches are typically replaced. Any and all rider input is carefully analyzed, which

Like fellow Frenchman Jean-Michel Bayle, Pichon won several championships (the 1995 and 1996 125cc Eastern Region Supercross), but was never totally accepted by American fans as a crowd favorite. Fortunately for Pichon, he wasn't booed like Bayle was during his first year stateside.

can sometimes result in drastic suspension, exhaust, and gearing changes. It's not like car racing where adjustments can be made during pit stops. From the moment the gate drops, there is nothing else that can be done from a mechanical standpoint; it is all up to the rider.

As time runs down, the riders usually take part in an autograph session, at which fans can

Who says that you have to be an American to have style? Multi-time World Champion Greg Albertyn was imported by Suzuki and has quickly adapted to the American style of riding.

TYPICAL SUPERCROSS TIMELINE

The day of a supercross is a hectic, full day, especially once the heat races begin!

8:00 A.M.	Tech inspection
11:00 P.M.	Riders' meeting
12:00 P.M.	Practice
3:00 P.M.	Qualifying races
7:00 P.M.	Opening ceremonies
7:30 P.M.	125cc heat race #1 (8 laps)
7:45 P.M.	125cc heat race #2
8:00 P.M.	250cc heat race #1
8:15 P.M.	250cc heat race #2
8:30 P.M.	125cc Last Chance Qualifier
8:45 P.M.	250cc Semi #1
9:00 P.M.	250cc Semi #2
9:15 P.M.	250cc Last Chance Qualifier
9:30 P.M.	125cc Main event (15 laps)
10:00 P.M.	250cc parade lap
10:15 P.M.	Main event (20 laps)
10:45 P.M.	Awards ceremony

David Pingree was a 1996 member of Mitch Payton's highly successful Pro Circuit team, which is run through Payton's shop. The program, which started on Hondas sponsored by Peak in the early 1990s, is now aboard SplitFire-backed Kawasakis.

RIGHT
Ezra Lusk put on a late-season charge at the conclusion of the 1996 race season and landed on the podium several times. His great rides may have been inspired by an Suzuki's lucrative bonus program that rewarded riders over $25,000 per victory. At the time, that was more than twice the going rate.

get complimentary team posters and have the opportunity to meet their heroes. It can be a magical moment when riders realize the true impact they have on other people's lives and a time when the adrenaline starts to flow.

Come dinnertime, the butterflies usually start to fly. Most meals are cooked in a microwave and are eaten as the riders watch videotape taken of their runs earlier during the day. This is the last opportunity to look for race lines and places to pass. Team managers also discuss strategy and team tactics when championships are on the line. If ever there were a time when racing could be viewed as a job, it is definitely now. After all, the teams wouldn't be sponsoring riders if there weren't money in it for them.

During opening ceremonies the hype really begins. A select few riders are introduced to the crowd, and the anticipation grows stronger. The other members of the field have to wait in the pits and get their first glimpse of the crowd when they enter the staging area.

Jeff Emig is always a crowd favorite no matter where the race is held. A fanatical group of Emig fans demonstrate their dedication by wearing fake affros to the races.

RIGHT

In 1995, Team Noleen garnered all kinds of attention thanks to its fleet of rocket-fast Yamahas. The manufacturer-backed team lost most of its steam in 1996, however, and all but slipped into extinction. Fortunately, it still served as a springboard for Kyle Lewis, one of the most under-rated and smooth riders on the circuit, who was recruited by Team Suzuki in 1997 to ride the Japanese National circuit.

Once racing begins, the object is to qualify for a spot in the main, and there are three ways that can be accomplished for 250cc riders. The first opportunity comes in one of two heat races where four riders from a group of 20 go straight to the main. From there, the remaining 16 go to one of two semis where the top five transfer. After that there is the Last Chance Qualifier, from which only the top two riders make it to the main. The rest have to watch the races from the stands.

The 125cc class is usually easier because transfer positions go deeper in the pack. The top 10 are placed into the final, and the remaining riders go directly into a take-two-to-the-main Last Chance Qualifier. There isn't a semi, but there are 22 riders in the 125cc main event, whereas there are only 20 in the 250cc class.

Prior to the main event, the riders are given a sighting lap. Since the track changes fairly rapidly during the night, this is a good opportunity to look for new lines and areas of the track that must be avoided; then it's show time.

After the main event, the top five riders are pulled into tech inspection where the bikes are gone over with a fine-tooth comb to guarantee that all the rules have been obeyed. In some instances, the AMA randomly selects machines for fuel tests, and it sometimes requires mechanics to disassemble entire engines to look for internal modifications that can be easily disguised by external appearance.

A Legendary Season

At the 1996 season kickoff in Orlando, Florida, it was commonplace to hear autograph seekers of all ages wish Jeremy McGrath luck. At the time, it seemed Jeremy could use it. Crowned as the three-time King of Supercross, the Honda factory rider was coming off knee surgery and had less then a month to prepare for his 1996 assault. It was only the second time in his career that he had been plagued by injury and the first such occasion that required surgery and rehabilitation.

But luck has rarely played a part in McGrath's career; he has always relied on sheer talent and a natural, relaxed air of confidence. The knee injury caused some to question his ability, and rumors spread that it could be the champ's last season.

In 1996, a southern California dealer formed Team Chaparral, and during its inaugural year the well-funded team nearly won the small-bore class at the Anaheim Supercross with Greg Schnell. Team Chaparral managed to race competitively against the factory bikes with production-based machines.

After missing the end of the supercross season with his broken jaw wired shut, Ryan Hughes had hoped to make up for his disappointing season by help-ing USA recover the team title at the annual Motocross des Nations in Jerez, Spain, but he suffered yet another physical setback. At the conclusion of the outdoor season, Hughes ran over his own foot and was once again benched. Jeff Emig would go to Europe in his place and bring gold back to the States.

On that warm January evening, McGrath served notice that the king was alive and well. McGrath clobbered the competition for the 29th time in his career, and even he was relieved. He chalked up wins in Dallas, Anaheim, Seattle, San Diego, Minneapolis, and so on. Injury wasn't going to stop McGrath, and soon many thought just the opposite. Was McGrath invincible?

After seven races, he set a new record for consecutive wins. He was decimating the compe-tition, and many of the other riders had commit-ted psychological suicide by conceding before the race had even begun. The race was for sec-ond, because everyone knew who would finish

first. Then came wins eight, nine, and ten. At this point, even people who were bored reading about McGrath's latest massacre in *Cycle News* began to take notice. Could the he win them all?

Rounds 11 and 12 were McGrath victories, but in round 13, McGrath's streak came to an end. Team Kawasaki's Jeff Emig stopped him in St. Louis and dislodged McGrath to second. It was the loss heard around the world. Ironically, both riders expressed relief over the result. Emig had finally broken through and won one, while McGrath was out of the pressure cooker. A week later, McGrath was back and won the season finale in Colorado by a typically vast margin of nearly 10 seconds.

LEFT

With total disregard for bike and body, Mike Metzger perfected a variety of aerial maneuvers. Metzger's tricks—the heel-clicker being the most notable—were featured on magazine pages and motocross videos, and Metzger assembled a following despite an entirely forgettable race record.

The race to the first turn is crucial in supercross. Passing is more difficult on the tight stadium tracks, and a mid-pack start gives the competition plenty of time to run away and hide. McGrath's number one Honda is taking the holeshot, as usual, while Windhom (38), Swink (23), Hughes (5), Jimmy Button (100), and the rest race for second place.

Winning 14 of 15 races in a season is a feat that may never be duplicated. In the competitive world of supercross, a poor start, a badly timed fall, or even a couple of trivial mistakes can take a rider off the podium. Passing is usually difficult on supercross tracks, and the terrain is unforgiving of errors.

By the close of the 1996 season, McGrath permanently etched his name into the record books and may have even elevated his popularity beyond that of Bob Hannah and Rick Johnson. But along the way, several other riders rose in popularity that year in both supercross and outdoors. From the beginning, McGrath had always stated that if he did lose an event, he expected it to come from his arch-rival Emig. Week in and week out, Emig was the one causing McGrath the most fits. The Team Kawasaki rider almost beat the Honda ace in San Diego during round five.

In the outdoor series, Emig's consistent second-place finishes put him in perfect position to pick up the pieces when McGrath injured his ankle at the end of the outdoor season. Hard work and a brilliant last-round ride rewarded Emig with the AMA 250cc National Championship.

McGrath, however, didn't get the competition he had originally anticipated from Emig's teammate, Damon Huffman except once during an incredible battle in Seattle. In the off-season, Huffman beat McGrath during the final round of the

Nevada's Casey Johnson has been brought up through Kawasaki's amateur program known as Team Green. The program acts as a factory farm system for the big leagues of professional racing.

World Supercross Championship in Geneva, Switzerland, in a heads-up race. Though Huffman was around during the first couple races of the AMA season, he too was a victim of a knee injury, but this one kept him sidelined for most of the season.

Ryan Hughes also played a pivotal role at the beginning of the year, but was struck down with a broken jaw in Charlotte, North Carolina. Ryan spent over a month with his jaw wired shut and was unable to maintain his normal training regimen.

Riding for an upstart team funded by Great Western Savings, Kawasaki support rider Phil Lawrence placed second in the opening round of the season and became a working-class hero overnight. Lawrence eventually turned out to be the highest-ranking non-factory rider of the season, and following in the tradition of Larry Ward, proved that you don't need a works bike to be competitive.

Surprisingly, McGrath didn't get much competition from the Honda camp. Steve Lamson injured his foot and sat out several stadium races so that he could concentrate on defending his AMA 125cc National Championship. That left the Honda of Troy squadron to pick up the slack for the powerful team if McGrath were to falter. In the preseason, the dealer-supported team run by Phil Alderton had acquired 1995 privateer sensation Larry Ward. The year before, Ward had placed second in the supercross series and boasted that he could beat McGrath if he were given works (read: unobtainable to the public) suspension. Honda gave it to him, but he didn't give Honda of Troy any wins. And neither did Mike Craig, Brian Swink, or Mike Brown, but the quartet did make the season flavorful.

The Yamaha camp had a pretty good arsenal of talent, as well. The year earlier, Damon Bradshaw decided to come out of self-imposed exile

Long-time factory Kawasaki rider Damon Huffman dominated the 125cc west division in the 1994 and 1995 supercross seasons. In 1996, he moved to the 250cc division with disappointing results brightened by one terrific ride. Big things are expected from Huffman in the future.

and wanted to give McGrath a run for his money. In the pre-McGrath era, Bradshaw was the rider to beat each and every week, yet he slipped into premature retirement without a single championship. If that wasn't bad enough, Bradshaw was off his mark more than some people, perhaps even he, himself, thought. Podium visits were few and far between. The same for teammate Doug

Henry, who spent most of the season trying to regain his strength and speed after a near deadly crash during the Budds Creek National in 1995. Both of the riders were still impressive at times, but it became obvious that some of their best days were behind them.

Team Suzuki didn't fare much better. After stealing Mike LaRocco from Kawasaki, the team run by

The Rewards of Speed

The sport of supercross offers a lucrative payoff for a few elite riders, and it is not uncommon to see teenage millionaires strolling the pits. A championship title can earn a rider more than $1 million, and a perfect season would be worth as much as $1.875 million.

The money comes in the form of factory salaries, which range from about $50,000 per year to ride for one of the smaller teams to several hundred thousand dollars to ride for one of the major manufacturer's teams (depending on the results and negotiation skill of the rider). Aftermarket companies will pay the factory to use certain components (handlebars, chains, oils, sprockets, graphics, and so on). The factory negotiates that contract and passes on the benefits to the rider, all of which is considered part of the factory salary.

Contracts with clothing manufacturers are also worth big bucks. Usually, the riders negotiate this contract, although Fox has had success signing entire teams to the same gear. Even so, the rider will negotiate with a helmet, boot, and goggle company separately. These clothing contracts are a big part of selling gear (who wouldn't want to wear the gear of their hero?) and are equally important to the riders. Riders have been known to switch teams just to stay with a particularly lucrative clothing contract.

A professional rider's other main source of revenue is bonuses. Every time a rider wins, he earns a check. Here is the approximate breakdown of how elite riders make their money:

```
Factory salary  . . . . . . . . . . . . . .$650,000

Clothing contract  . . . . . . . . . . .$200,000
(Pants, jersey, gloves, and kidney belt)
Helmet  . . . . . . . . . . . . . . . . . .$100,000
Boots . . . . . . . . . . . . . . . . . . . .$75,000
Goggles  . . . . . . . . . . . . . . . . . .$30,000
Gear Total . . . . . . . . . . . . . . . . .$405,000

Bonuses
250cc SX win bonus
($12,000 each x 15 races)  . . . . .$180,000
Championship bonus
($120,000 each x 2 titles)  . . . . .$240,000
AMA Racing Season Total  . .$1,475,000

Post-season racing overseas
(Japan and Europe)  . . . . . . . .$400,000
Grand Total  . . . . . . . . . . . . .$1,875,000
```

five-time World Champion Roger DeCoster suffered major teething problems with its all-new Honda-like RM250. Mechanical failures were occurring all too often, and rider confidence was down. LaRocco never looked comfortable with the bike the entire season, and two-time World Champion Greg Albertyn spent most of his season on the ground.

If Suzuki had a hope, it was Ezra Lusk. The talented Georgian could match McGrath's pace on occasion, but suffered from poor starts like his counterpart LaRocco. At one event Lusk charged from dead last to finish runner-up in the Pontiac Silverdome.

At the end of the season, McGrath's father expressed relief that Jeremy hadn't won it all. "He needs something to shoot for each year," Jack McGrath said. "Every year people say it can't get any better, and Jeremy proves them wrong each time. Had he won every race this season, it would have been downhill from here on out, unless he continued to go unbeaten. I would have liked that, but I don't think the fans would have appreciated it too much."

Whatever the future brings for McGrath, he will continue to be the supercross story of the 1990s, and perhaps for much longer than that.

Steve Lamson helped Honda have what was essentially a near-perfect team. Along side of Jeremy McGrath, Lamson is always a contender in supercross and has twice won the 125cc Outdoor National Championship.

THE POINTS SYSTEM

Every form of racing has its own point system. In NASCAR, a driver can earn championship points for obtaining the pole, leading a lap, leading the most laps, and, of course, winning. In a given day, 100 or so points can be won or, worse, lost.

Point systems in every sport are always controversial, and the method of calculation oftentimes can change the outcome of a championship. In 1992, Damon Bradshaw lost the 250cc Supercross title after winning nine races in the 15-race series. Many of his fans and even some of his competitors felt that he was robbed. The argument always involves consistency versus winning, and which should reward riders the greatest. In that same 1992 season, Jeff Stanton only won three races, but his consistency landed him the championship.

In contrast to NASCAR, there are no bonus points in supercross. Instead, points are simply awarded by finish position in the main event (or in Outdoor Nationals, each moto). The points are awarded as follows:

Position	Points
1st	25
2nd	22
3rd	20
4th	18
5th	16
6th	15
7th	14
8th	13
9th	12
10th	11
11th	10
12th	9
13th	8
14th	7
15th	6
16th	5
17th	4
18th	3
19th	2
20th	1

Mike LaRocco made a daring career decision when he quit Team Kawasaki and signed with Suzuki in 1996. LaRocco is one of the most dedicated professionals the sport has ever produced, but his career has been plagued by injuries that have severely limited his potential.

By all accounts, John Dowd is a late bloomer. The New Englander got his first factory ride at age 30 and then shocked the world by challenging for the 125cc Eastern Region title in 1996. During that season, Dowd rewarded Yamaha with second in the series championship.

For first time since 1982 (Mark Barnett), a Suzuki carried the number one plate in the 250cc class. In a surprising late season move, Jeremy McGrath signed to ride Suzukis for 1997. Racing as Team Nac Nac, Jeremy rode a factory Suzuki and worked with mechanic Wyatt Seals out of Team Suzuki's transporter.

CHAPTER 6

LIFE AFTER SUPERCROSS

When the supercross season wraps up, usually in May or June, a professional racer's season is just getting into full swing. The riders switch immediately into the AMA Outdoor National Series, which runs into early fall, and then have the option of tackling a number of race series that take place outside the regular AMA season.

Born in Europe, true motocross is worlds apart from the ultra-clean, almost surreal world of stadium racing. It doesn't have to abide by the normal space restrictions that track builders have to cope with in supercross. For starters, the race courses are composed of natural terrain, which oftentimes includes drastic elevation changes natural to the region. The tracks are also longer, produce higher speeds, and generally have closer racing. To purists, motocross is the lifeblood of the sport, and they actually prefer it to supercross. It's intense, and as you will find out, it can be pretty wild.

While supercross races reward precision riding over tricky, technical obstacles, outdoor motocross racing requires brute strength, intense conditioning, and gutsy riding. The races are longer, both in time and distance, and motos typically last about 30 minutes. As the race drags on, even the best-conditioned riders fade and begin to make mistakes, and a racer that comes into the race slightly out-of-shape will struggle to finish in a respectable fashion.

Outdoor motocross courses are changing to incorporate more supercross-style obstacles (mainly big double or triple jumps), but the courses are still laid out on natural terrain.

Monster jumps aren't exclusive to supercross, as Jeremy McGrath demonstrates by airing it out over Mike LaRocco.

Battling the elements is a crucial aspect of outdoor motocross racing. Here, Chad Pederson tries to beat the heat with holes torn in his jersey for venting. Despite a solid 1996 season, Pederson was dropped by Pro Circuit and switched to the Arenacross series the next year.

Professional racers contest two classes in the AMA's outdoor series, the 125 and 250cc classes. While the 250cc class is definitely the premier class, the 125cc class gets more respect than it does in supercross, where 125cc racing is considered a proving ground for younger up-and-coming riders.

As in supercross, lower-level riders must ride in qualifying heats to make the main event. Higher-ranked riders, however, do not have to qualify. Once qualifying is over, racers in the main event ride two motos. Points are awarded for each moto, and an overall winner is declared, with more weight given to the second moto finish.

For example, If Ted finishes first in the first moto and second in the second moto, the results would be tallied as finishing 1-2. If Bob finished 2-1, the second moto would break the tie, and Bob would win the overall.

Bike setup for outdoor motocross racing is different than supercross, although it varies from track to track. Outdoor tracks generally favor bikes with lots of horsepower. Tractability becomes less of a factor on open, fast tracks with straights that allow the rider to hold the

Mud is much more common on the outdoor circuit. As in supercross, the slop and slime is the great equalizer. Ryan Hughes roosts his way through the muck on the SplitFire KX125 at Mt. Morris in 1995.

Steve Lamson was truly in a class of his own during the 1996 season as he successfully defended his AMA 125cc National Championship. He won 9 of the first 11 races to clinch the championship with two rounds remaining. His only true rival was Kevin Windham, the only other 125cc rider to take the overall win at an Outdoor National in 1996.

throttle open. Supercross tracks favor bikes with instantaneous power, which means low-end grunt. The wheels of the bike are only on the ground for a few seconds, and the rider needs to build momentum as quickly as possible to clear giant triples or tricky combination jumps. Suspension settings and gearing are also altered for outdoor motocross, and these settings vary slightly from track to track.

From a fan's point of view, watching an outdoor race is quite different as well. On most motocross tracks, you can't really see the whole course, meaning you might not see every pass or crash. The advantage is that you can get up close and personal.

Other Race Series

At the conclusion of the AMA racing season, it would be natural to assume that the riders would take a well-deserved three-and-a-half month break. After all, they have spent almost 30 weekends racing straight through September with only a handful of days off. In reality, though, a top racer never has any chance to rest. At best, they may slip in a couple days at the river riding personal watercraft or water skiing, but most of them head to Europe or the Orient to spend the winter racing.

For the last decade, the international race scene in countries like France, Spain, Switzerland, Germany, Italy, and Japan have turned out to be extremely lucrative for racers. Riders like Jeremy McGrath can make about $400,000 in a few short weeks and are guaranteed start money. In fact, McGrath commands $100,000 in start money for the three-day supercross in France; not bad for a weekend's work!

This is a sharp contrast to racing in the United States. At home, most supercrosses only pay $5,000 to win (the exception is usually Anaheim and Daytona, which pay the victor $10,000 because of higher attendance). Sound illogical? Well it is, sort of. You see, in addition to higher ticket prices overseas (around $50-$75), foreign promoters are able to sell television rights to networks for big money, which translates into more prize money. After all, some motorsport categories are bigger overseas and therefore it is not uncommon to see supercross live on television.

In 1996, John Dowd was 32 years old and hammering away on his factory Yamaha, competing against riders who are typically a decade younger. His determination triumphs over Father Time, and he is an inspiration to vet racers everywhere.

NEXT PAGES
The higher speeds associated with the outdoor nationals make for spectacular first-turn action. Considering that 40 riders vie for the front spot in a line typically wide enough for three or four bikes, the rarity of major first-turn pile-ups is a testament to these rider's skill.

73

Honda experimented with a factory-supported race team run through Bill's Pipes in 1996, but the team was plagued by bad luck and injury that mostly centered around teen sensation Robbie Reynard of Oklahoma.

EuroSport, Europe's version of ESPN, tapes the heat races, semis, and Last Chance Qualifiers, and then broadcasts the 125cc and 250cc main events live in over two dozen countries.

Another reason for the higher payoff is the fact that most of the talent pool resides in the United States. Though France has produced heroes like Jean-Michel Bayle, Mickael Pichon, and Yves DeMaria, most of the global media attention is focused on the U.S. stadium series and the American riders who dominate indoors. It also doesn't hurt that U.S. riders have proven their dominance by winning 14 of the past 16 titles at the annual Motocross des Nations, an

event where each country sends its three best riders to seek world-wide acclaim in a three moto, go-for-broke, put-your-reputation-on-the-line championship.

The lure of the big money is the driving force that gets riders and their mechanics to make the 7,000- to 10,000-mile journey. By no

RIGHT
Team Honda has an unparalleled nose for talent, along with a knack for hiring riders at the right time. Scott Sheak's great showing as a privateer in 1996 earned him a spot on the 1997 Team Honda squad. One of the pluses for Sheak will be the team's new half-million-dollar semi.

One of the lures of outdoor racing is that it puts the action so close to the spectators. Here at Steel City, the fans were able to get within a few feet of the down-to-the-wire championship duel between Jeff Emig and Jeremy McGrath.

means is this just another race on the schedule. It's cost-prohibitive for riders to have their team semis shipped to Europe and Japan, so the riders and their mechanics have to consolidate their racing effort and check all of their racing equipment on planes as part of their normal luggage. This includes works suspension, high-performance cylinders, cylinder heads, special carburetor bodies, hand-built exhaust systems, magnesium triple clamps, special-bend handle-

bars, works tires, different sprockets and skid plates, team-issue graphics and seat covers, riding gear, and of course, all the tools to install the vast array of high-tech equipment. The bikes themselves are supplied by various distributors.

RIGHT
After falling in the famous Southwick sand during the 1996 National, Suzuki's Tim Ferry found out what it's like to be sand-blasted at close range.

Doug Henry won the 1994 125cc Outdoor National Championship for Team Honda. Riding the 250cc class for them in 1995, he broke his back when he skyshot a drop-off jump at Budds Creek. Most said his career was over, but Henry defied the odds and made a historic comeback in 1996, signing with the Yamaha factory team. He picked up speed throughout the season, and a moto win at Washougal proved that he regained his speed.

When you continually test your limits, you are bound to exceed them every once in a while as Ezra Lusk did here. Motocross is a physically demanding sport, and crashes are all too common.

At most events, the upper echelon of riders are loaned new machinery, which their mechanics transform into semi-works bikes overnight. The other riders are given slightly used bikes on which to compete.

Come show time, the events themselves are more exotic and entertaining than stateside supercrosses because there is much more pageantry. Riders are treated like gods, worshipped like kings, and are true celebrities in every facet of the imagination. Mini-rock con-

certs, indoor fireworks shows, wild jump contests, and spectacular laser light shows attract sell-out crowds. In fact the 15,000-seat OmniSports complex in Paris sells out three consecutive days, and ticket scalpers are able to get as much as three times the face value!

But what the fans really pay to see is racing, and they are rewarded with some of the best two-wheel action on the planet. Several of the overseas events are part of the World Supercross Series which continues to gain momentum each and every year since its inception in the early 1990s. While it's nowhere near as prestigious as the AMA Supercross Series, it's still a title that every rider would like to add to their resume.

If you have ever played chicken, you can imagine what it's like to attack a first turn at a national-caliber pace. Back off the throttle first and you lose a very intense drag race.

Usually by the end of November, most of the factory riders are forced by their factory contract to return home to finish development on next year's racing machinery. The true testing period for research and development of factory equipment is compacted into four to eight very intense weeks. Research and development runs throughout the season, but the bulk of the development is done during the months of November, December, and January. This time is when the race teams are handed production machinery on which to experiment, and they work furiously to get the new bikes set up properly for their riders, hopefully with a bit of an edge over the competition.

The testing is predominantly done in southern California, which is home for the corporate headquarters of Honda, Kawasaki, Suzuki , and Yamaha. Each of these teams lease property and have private supercross tracks constructed for testing and practicing purposes. Access to these tracks and the intense testing that takes

Jeremy McGrath, uncharacteristically out of shape as he hangs it all out trying to catch his friend and nemesis Jeff Emig in the final round of the 1996 250cc Outdoor National Series. The winner would take home the championship and, for once, it wouldn't be Jeremy.

Jeff Emig capped off a consistent 1996 Outdoor National Series with perhaps the best ride of his life at the final round at Steel City, where he clearly out-rode the fastest man on the planet, Jeremy McGrath, to take home the 250cc Outdoor National Championship.

While outdoor nationals can't accommodate the huge numbers of fans that can be packed into stadiums, attendance averages over 10,000 folks. That number has been steadily on the rise in the mid-1990s, probably due to the sport's increased exposure due to supercross and expanded television coverage on ESPN and ESPN2.

place gives factory riders a huge advantage over their privateer competition.

The testing itself is usually handled by a vast array of highly trained technicians. First, each and every idea has to pass a brainstorm session where engineers discuss the pros and cons of the various changes. Once a decision has been made to test the product, it is turned over to the machine shop where it's fabricated using exotic materials that maximize strength and minimize weight.

Once built, every product has to undergo a series of evaluations against the stop watch and other race bikes. Many teams are also experimenting with computerized engine and suspension telemetry that helps determine effectiveness. Of course, rider input is highly critical, but it's not uncommon for riders to feel that something is faster when in actuality it is slower. Mellow powerbands feel slower but may yield faster lap times due to less wheelspin. Riders often rely on stop watches and videotapes to measure the worth of a new development.

The final deciding factor is reliability. When the average factory race team spends approximately $3,5000,000 on its racing effort, mechanical failure is unacceptable. Most companies insist that a product must produce 20 problem-free-hours of performance before it can enter the racing arena. Durability is so important that most products don't actually have to show any performance gain whatsoever, provided that it wards off equipment failure. That's why most race bikes have thicker spokes, special hubs and triple clamps, and added frame gusseting.

The engine, suspension, chassis, tire, and ergonomic testing have to be conducted simultaneously because of time constraints. Though this may sound chaotic, the engineers usually pool information and co-develop products among testing divisions. The best part of all of this is that although the technology is developed to give the race team a performance advantage, the things that work usually show up in dealer showrooms, where they help consumers fulfill racing dreams of their own.

RIGHT
Yamaha's Kevin Windham won the last two races of the 1996 season and was definitely the closest anyone got to the seemingly invincible Steve Lamson. All season long, Windham had been working under Bob Hannah's tutelage, and the hard work was very apparent.

The eyes of a champion.

For three consecutive nights, fans pour inside the 15,000-seat stadium in Bercy, France, for the most spectacular supercross show in the world. One of the things that distinguishes the races on the two continents are the jump contests. It's during these contests that some of the most famous mid-air moves were created, and some of them are not allowed at AMA-sanctioned events. This is Jeremy McGrath in action.

NEXT PAGES
The Paris Supercross kicks off with a spectacular light show while rock music fills the air. Notice that the Europeans still use over and under bridges, something that hasn't been seen in the United States in almost a decade. The event is broadcast live on EuroSport, Europe's equivalent to ESPN.

RIGHT

In drastic contrast to the overseas supercross scene, riders selected to represent Team USA in the Motocross des Nations do so for national pride rather than money. Once a joke in international competition, Team USA became the dominant force in the 1980s and early 1990s. After winning 13 times in a row, Team USA finished second in the 1994 and 1995. The 1996 team of Jeremy McGrath, Steve Lamson, and Jeff Emig dominated, and brought the title home. Lamson, fired up to avenge a 1995 MXdN drubbing by Sebastian Tortelli pulled away from Tortelli and the rest of the pack to become the first rider to win an overall on a 125cc bike.

Jeremy McGrath proved to be the fastest rider in the world after winning both of his motos during the 1996 Motocross des Nations in Jerez, Spain. It was one of the rare occasions that McGrath had the opportunity to race against international riders on an outdoor circuit, and he responded by winning both of his motos by a wide margin.

Big-bore two-strokes like this KX500 have been extinct on the professional U.S. racing scene since 1993, but Jeff Emig still managed to man-handle the beast to help USA win for the 14th time in the last 16 attempts. Ryan Hughes rode the powerful Kawasaki in 1995 (with Emig on a 250) and was slated to return in 1996, but a late-season injury kept Ryno back in the states for the MXdN race.

APPENDIX
STATISTICS, WIN STREAKS AND RECORDS

The Jeremy File

- Jeremy McGrath is the all-time supercross win leader with 43 victories as of December 1996.
- Jeremy McGrath has entered 61 supercross races since 1993 and has won 70 percent of them, losing only 18 times. Prior to that, he only raced four 250cc supercross races when his 125cc Western Region Supercross series was on hiatus.
- Jeremy McGrath holds the record for most consecutive supercross wins at 13, set in 1996.
- Jeremy McGrath holds the record for the most supercross wins in a single season. In 1996, he won 14 of 15 races. Jeff Emig's win at the St. Louis Supercross was McGrath's only supercross loss that year.
- Counting his two 125cc Western Region Supercross Championships (1991–92), Jeremy McGrath has won six straight Supercross Championships.

King of the Track

- Jeff Stanton has won the Daytona Supercross four times in a row (1989–92). Mike Kiedrowski is a close second in this category with three wins at Daytona (1993–95).
- Jeff Ward won the 500cc class at Steel City in Delmont, Pennsylvania, a record four consecutive times (1989–92).
- Jeff Stanton won the 250cc class in Southwick, Massachusetts, a record four consecutive times (1989–92).

- Jeff Matiasevich won the 125cc Western Region Supercross in Seattle four times in a row (1988–89). In those two years, Seattle offered two nights of racing back-to-back.
- Rick Johnson won the Seattle Supercross six out of seven times from 1986–89. Ron Lechien beat him on February 14, 1988.
- Bob Hannah won the Pontiac Supercross in Pontiac, Michigan, six consecutive times from 1977–79. The venue was a doubleheader until 1994.
- Jeremy McGrath won his hometown supercross (Anaheim, California) race six times in six years if you count his two 125cc Western Region victories.
- Jeremy McGrath has won all of the supercross races held at the Metrodome in Minneapolis, Minnesota (four races from 1993 to 1996).

Consecutive Wins

- Ake Johnsson won 27 consecutive 500cc Trans-AMA motos in 1972.
- Brad Lackey won six consecutive 500cc Nationals in 1972.
- Bob Hannah won eight consecutive races in the 1978 250cc National Series. He lost the final two races to Kent Howerton and Jimmy Ellis.
- Bob Hannah won four consecutive races in the 1979 250cc National Championship.
- Broc Glover won four consecutive races in the 1983 125cc National Championship.

- Bob Hannah won six consecutive supercrosses in 1978. Rick Johnson tied Bob Hannah's record of six consecutive supercrosses 10 years later in 1988.
- Jeremy McGrath more than doubled the record when he pulled off 13 straight victories in 1996.

Domination
- Marty Smith won all but one of the 125cc Nationals in 1975. He lost the opener on April 6 to Tim Hart but went on to win the remaining six races (and all 12 motos) in the series.
- Bob Hannah won a record 27 supercrosses from March 5, 1977, to March 9, 1985. A few years later Rick Johnson broke that record and ended his career with 28 victories. As of press time, Jeremy McGrath continues to build upon his new record of 43.
- In 1980, Kent Howerton won six of seven races to give Suzuki the 250cc National Championship.
- Mark Barnett won the first 14 motos of the 1981 125cc Outdoor Nationals.
- In 1981, Broc Glover won 8 of 10 500cc Nationals. Chuck Sun beat him twice and both defeats came back to back.
- Jeff Ward won 8 out of 10 125cc Nationals on his way to the 1984 Championship.
- David Bailey won 8 of 10 500cc Nationals on his way to the 1984 title.
- Rick Burgett won six out of seven 500cc Nationals in 1978. Rex Staten beat him on August 13.

- Jeff Emig won six of seven consecutive races during the 1992 125cc National Championship Series. That year there were a total of 11 races.
- Mike Kiedrowski won six of seven races during the 1993 250cc National Championship Series. Damon Bradshaw beat him in Pennsylvania on May 30.
- Jeremy McGrath made the podium every single time in the 1996 supercross season.
- Steve Lamson won the 9 of the first 11 Nationals on his way to the 1996 125cc Championship.
- Team USA leads in most consecutive Motocross des Nations and Trophy des Nations wins at 13 (1981–93).
- Team Honda won nine consecutive Supercross Championships starting with Rick Johnson in 1988 to Jeremy McGrath in 1996.

Amazing Feats
- In 1995, Steve Lamson overcame a 63-point deficit to win the 125cc National Championship.
- Damon Bradshaw has won the most supercrosses without winning the championship. He won nine main events in 1992.
- In 1989, Mike Kiedrowski won the 125cc National Championship, a year after he lost the title by just one point.
- Ron Lechien is the youngest rider to win a 250cc supercross. He won Orlando, Florida, aboard a Yamaha YZ250 at age 16 on June 11, 1983.

250cc Supercross

1996 Jeremy McGrath (Hon)
1995 Jeremy McGrath (Hon)
1994 Jeremy McGrath (Hon)
1993 Jeremy McGrath (Hon)
1992 Jeff Stanton (Hon)
1991 Jean-Michel Bayle (Hon)
1990 Jeff Stanton (Hon)
1989 Jeff Stanton (Hon)
1988 Rick Johnson (Hon)
1987 Jeff Ward (Kaw)
1986 Rick Johnson (Hon)
1985 Jeff Ward (Kaw)
1984 Johnny O'Mara (Hon)
1983 David Bailey (Hon)
1982 Donnie Hansen (Hon)
1981 Mark Barnett (Suz)
1980 Mike Bell (Yam)
1979 Bob Hannah (Yam)
1978 Bob Hannah (Yam)
1977 Bob Hannah (Yam)
1976 Jim Weinert (Kaw)
1975 Jim Ellis (Can-Am)
1974 Pierre Karsmakers (Yam)

500cc Supercross

1975 Steve Stackable(Maico)
1974 Gary Semics (Hus)

125cc Eastern Region Supercross

1996 Mickael Pichon (Kaw)
1995 Mickael Pichon (Kaw)
1994 Ezra Lusk (Suz)
1993 Doug Henry (Hon)
1992 Brian Swink (Suz)
1991 Brian Swink (Suz)
1990 Denny Stephenson (Suz)
1989 Damon Bradshaw (Yam)
1988 Todd DeHoop (Suz)
1987 Ron Tichenor (Suz)
1986 Keith Turpin (Suz)
1985 Eddie Warren (Kaw)

125cc Western Region Supercross

1996 Kevin Windham (Yam)
1995 Damon Huffman (Suz)
1994 Damon Huffman (Suz)
1993 Jimmy Gaddis (Kaw)

1992 Jeremy McGrath (Hon)
1991 Jeremy McGrath (Hon)
1990 Ty Davis (Hon)
1989 Jeff Matiasevich (Kaw)
1988 Jeff Matiasevich (Kaw)
1987 Willie Surratt (Suz)
1986 Donny Schmit (Suz)
1985 Bobby Moore (Suz)

500cc Outdoor National

1993 Mike LaRocco (Kaw)
1992 Mike Kiedrowski (Kaw)
1991 Jean-Michel Bayle (Hon)
1990 Jeff Ward (Kaw)
1989 Jeff Ward (Kaw)
1988 Rick Johnson (Hon)
1987 Rick Johnson (Hon)
1986 David Bailey (Hon)
1985 Broc Glover (Yam)
1984 David Bailey (Hon)
1983 Broc Glover (Yam)
1982 Darrell Schultz (Hon)
1981 Broc Glover (Yam)
1980 Chuck Sun (Hon)
1979 Danny LaPorte (Suz)
1978 Rick Burgett (Yam)
1977 Marty Smith (Hon)
1976 Kent Howerton (Hus)
1975 Jim Weinert (Yam)
1974 Jim Weinert (Kaw)
1973 Pierre Karsmakers (Yam)
1972 Brad Lackey (Kaw)
1971 Mark Blackwell

250cc Outdoor National

1996 Jeff Emig (Kaw)
1995 Jeremy McGrath (Hon)
1994 Mike LaRocco (Kaw)
1993 Mike Kiedrowski (Kaw)
1992 Jeff Stanton (Hon)
1991 Jean-Michel Bayle (Hon)
1990 Jeff Stanton (Hon)
1989 Jeff Stanton (Hon)
1988 Jeff Ward (Kaw)
1987 Rick Johnson (Hon)
1986 Rick Johnson (Hon)
1985 Jeff Ward (Kaw)

1984 Rick Johnson (Hon)
1983 David Bailey (Hon)
1982 Donnie Hansen (Hon)
1981 Kent Howerton (Suz)
1980 Kent Howerton (Suz)
1979 Bob Hannah (Yam)
1978 Bob Hannah (Yam)
1977 Tony DiStefano (Suz)
1976 Tony DiStefano (Suz)
1975 Tony DiStefano (Suz)
1974 Gary Jones (Hon)
1973 Gary Jones (Hon)
1972 Gary Jones (Yam)
1971 Gary Jones

125cc Outdoor National

1996 Steve Lamson (Hon)
1995 Steve Lamson (Hon)
1994 Doug Henry (Hon)
1993 Doug Henry (Hon)
1992 Jeff Emig (Yam)
1991 Mike Kiedrowski (Kaw)
1990 Guy Cooper (Suz)
1989 Mike Kiedrowski (Hon)
1988 George Holland (Hon)
1987 Mickey Dymond (Hon)
1986 Mickey Dymond (Hon)
1985 Ron Lechien (Hon)
1984 Jeff Ward (Hon)
1983 Johnny O'Mara (Hon)
1982 Mark Barnett (Suz)
1981 Mark Barnett (Suz)
1980 Mark Barnett (Suz)
1979 Broc Glover (Yam)
1978 Broc Glover (Yam)
1977 Broc Glover (Yam)
1976 Bob Hannah (Yam)
1975 Marty Smith (Hon)
1974 Marty Smith (Hon)

American World Champions

1982 Danny LaPorte (Yamaha 250)
1982 Brad Lackey (Suzuki 500)
1989 Trampas Parker (KTM 125)
1991 Trampas Parker (Honda 250)
1990 Donny Schmit (Suzuki 125)
1992 Donny Schmit (Yamaha 250)
1994 Bobby Moore (Yamaha 125)

American Motocross des Nations and Trophy des Nations Results

1996 Steve Lamson, Jeremy McGrath, Jeff Emig (1st)

1995 Steve Lamson, Jeff Emig , Ryan Hughes (2nd)

1994 Jeff Emig, Mike Kiedrowski, Mike LaRocco (2nd)

1993 Jeff Emig, Jeremy McGrath, Mike Kiedrowski (1st)

1992 Billy Liles, Mike LaRocco, Jeff Emig (1st)

1991 Mike Kiedrowski, Jeff Stanton, Damon Bradshaw (1st)

1990 Jeff Ward, Jeff Stanton, Damon Bradshaw (1st)

1989 Jeff Ward, Jeff Stanton, Mike Kiedrowski (1st)

1988 Ron Lechien, Rick Johnson, Jeff Ward (1st)

1987 Bob Hannah, Rick Johnson, Jeff Ward (1st)

1986 David Bailey, Rick Johnson, Johnny O'Mara (1st)

1985 David Bailey, Jeff Ward, Ron Lechien (1st)

1984 Rick Johnson, Johnny O'Mara, Jeff Ward, David Bailey (1st)

1983 David Bailey, Mark Barnett, Jeff Ward, Broc Glover (1st)

1982 David Bailey, Johnny O'Mara, Danny LaPorte, Donnie Hansen (1st)

1981 Chuck Sun, Johnny O'Mara, Danny LaPorte, Donnie Hansen (1st)

1980 U.S. didn't field a team

1979 U.S. didn't field a team

1978 Bob Hannah, Rick Burgett, Chuck Sun, Tommy Croft (4th)

1977 Kent Howerton, Tony DiStefano, Gary Semics, Steve Stackable (2nd)

1976 Rex Staten, Bob Hannah, Tony DiStefano, Kent Howerton (5th)

1975 Jim Pomeroy, Brad Lackey, Kent Howerton, Tony DiStefano (9th)

1974 Jim Pomeroy, Brad Lackey, Tony DiStefano, Jim Weinert (2nd)

1996 Supercross Final Point Standing

250cc

1.	Jeremy McGrath	372
2.	Jeff Emig	240
3.	Ezra Lusk	215
4.	Ryan Hughes	202
5.	Phil Lawrence	202
6.	Mike LaRocco	200
7.	Damon Bradshaw	187
8.	Larry Ward	178
9.	Damon Huffman	172
10.	Brian Swink	156

125cc Eastern Region

1.	Mickael Pichon	232
2.	John Dowd	207
3.	Nathan Ramsey	167
4.	Scott Sheak	137
5.	Brian Deegan	136
6.	Tim Ferry	128
7.	Cory Keeney	109
8.	Davey Yezek	106
9.	Jeff Willoh	46
10.	Rusty Holland	41

125cc Western Region

1.	Kevin Windham	170
2.	James Dobb	127
3.	Kim Ashkenazi	99
4.	Greg Schnell	97
5.	Michael Brandes	89
6.	Casey Johnson	85
7.	Pedro Gonzalez	73
8.	Jeff Willoh	58
9.	David Pingree	53
10.	David Vuillemin	45

1996 Outdoor National Point Standings

125cc National

1.	Steve Lamson	584
2.	Kevin Windham	455
3.	John Dowd	439
4.	Buddy Antunez	301
5.	Mike Craig	299
6.	Tim Ferry	271
7.	Ezra Lusk	259
8.	Robbie Reynard	244
9.	Chad Pederson	244
10.	James Dobb	243

250cc National

1.	Jeff Emig	566
2.	Jeremy McGrath	556
3.	Mike LaRocco	428
4.	Greg Albertyn	403
5.	Larry Ward	350
6.	Kyle Lewis	306
7.	Brian Swink	300
8.	Ryan Hughes	297
9.	Damon Bradshaw	297
10.	Jimmy Button	272

Statistics compiled November 1996.

RIVERDALE PUBLIC LIBRARY DISTRICT

INDEX